CHEMICAL TRANSFER OF LEARNED INFORMATION

NORTH-HOLLAND RESEARCH MONOGRAPHS

FRONTIERS OF BIOLOGY

VOLUME 22

Under the General Editorship of
A. NEUBERGER
London
and
E. L. TATUM
New York

NORTH-HOLLAND PUBLISHING COMPANY
AMSTERDAM · LONDON

CHEMICAL TRANSFER OF LEARNED INFORMATION

Edited by

EJNAR J. FJERDINGSTAD

Institute of General Zoology, University of Copenhagen
Copenhagen, Denmark

1971

NORTH-HOLLAND PUBLISHING COMPANY
AMSTERDAM·LONDON

Library of Congress Catalog Card Number: 73–166308
ISBN North-Holland: 0 7204 7122 2
ISBN American Elsevier: 0 444 10112 8

51 graphs, 56 tables, 2 illustrations.

PUBLISHERS:

NORTH-HOLLAND PUBLISHING COMPANY - AMSTERDAM
NORTH-HOLLAND PUBLISHING COMPANY, LTD. - LONDON

SOLE DISTRIBUTORS FOR THE U.S.A. AND CANADA:

AMERICAN ELSEVIER PUBLISHING COMPANY, INC.
52 VANDERBILT AVENUE
NEW YORK, N.Y. 10017

PRINTED IN THE NETHERLANDS

Previous volumes in the series FRONTIERS OF BIOLOGY

Volume 1:	**Microbial models of cancer cells** G. F. GAUSE, Academy of Medical Sciences, Moscow
Volume 2:	**Interferons** Edited by N. B. FINTER, Imperial Chemical Industries, Macclesfield, Cheshire
Volume 3:	**The biochemical genetics of vertebrates except man** I. E. LUSH, Rowett Research Institute, Aberdeen
Volume 4:	**Delayed hypersensitivity** J. L. TURK, University of London
Volume 5:	**Human population cytogenetics** W. M. COURT BROWN, Western General Hospital, Edinburgh
Volume 6:	**Thymidine metabolism and cell kinetics** J. E. CLEAVER, University of California
Volume 7:	**The cell periphery, metastasis and other contact phenomena** L. WEISS, Roswell Park Memorial Institute, Buffalo, N.Y.
Volume 8:	**The electrostatics of biological cell membranes** R. M. FRIEDENBERG, University of Maryland
Volume 9:	**The pyrrolizidine alkaloids. Their chemistry, pathogenicity and other biological properties** L. B. BULL, C. C. J. CULVENOR and A. T. DICK, Commonwealth Scientific & Industrial Research Organization, Australia
Volume 10:	**Antagonists and nucleic acids** M. EARL BALIS, Sloan Kettering Institute for Cancer Research, Rye, N.Y.
Volume 11:	**Macrophages and immunity** D. S. NELSON, Department of Bacteriology, The University of Sydney, Australia
Volume 12:	**The biological code** M. YČAS, Department of Microbiology, State University of New York, N.Y.
Volume 13:	**The biochemistry of folic acid and related pteridines** R. L. BLAKLEY, Department of Biochemistry, John Curtin School of Medical Research, Australian National University
Volume 14:	**Lysosomes in biology and pathology** (complete in 2 volumes) Edited by J. T. DINGLE and HONOR B. FELL, Strangeways Research Laboratory, Cambridge
Volume 15:	**Handbook of molecular cytology** Edited by A. LIMA-DE-FARIA, Institute of Molecular Cytogenetics, University of Lund, Sweden
Volume 16:	**Immunogenetics of tissue transplantation** ALENA LENGEROVÁ, Institute of Experimental Biology and Genetics, Czechoslovak Academy of Sciences, Prague
Volume 17:	**Prenatal respiration** HEINZ BARTELS, Physiologisches Institut der Medizinischen Hochschule, Hannover, Germany
Volume 18:	**The transmission of passive immunity from mother to young** F. W. ROGERS BRAMBELL, Emeritus Professor of Zoology, Bangor, Caerns., United Kingdom
Volume 19:	**The principles of human biochemical genetics** HARRY HARRIS, Galton Laboratory, University College, London
Volume 20:	**Foetal and neonatal immunology** J. B. SOLOMON, Department of Bacteriology, The University of Aberdeen, Scotland
Volume 21:	**Hemoglobin and myoglobin in their reactions with ligands** E. ANTONINI and M. BRUNORI, Center of Molecular Biology of C.N.R., Institute of Biochemistry and Regina Elena Institute for Cancer Research, Rome

General preface

The aim of the publication of this series of monographs, known under the collective title of '*Frontiers of Biology*', is to present coherent and up-to-date views of the fundamental concepts which dominate modern biology.

Biology in its widest sense has made very great advances during the past decade, and the rate of progress has been steadily accelerating. Undoubtedly important factors in this acceleration have been the effective use by biologists of new techniques, including electron microscopy, isotopic labels, and a great variety of physical and chemical techniques, especially those with varying degrees of automation. In addition, scientists with partly physical or chemical backgrounds have become interested in the great variety of problems presented by living organisms. Most significant, however, increasing interest in and understanding of the biology of the cell, especially in regard to the molecular events involved in genetic phenomena and in metabolism and its control, have led to the recognition of patterns common to all forms of life from bacteria to man. These factors and unifying concepts have led to a situation in which the sharp boundaries between the various classical biological disciplines are rapidly disappearing.

Thus, while scientists are becoming increasingly specialized in their techniques, to an increasing extent they need an intellectual and conceptual approach on a wide and non-specialized basis. It is with these considerations and needs in mind that this series of monographs, '*Frontiers of Biology*' has been conceived.

The advances in various areas of biology, including microbiology, biochemistry, genetics, cytology, and cell structure and function in general will be presented by authors who have themselves contributed significantly to these developments. They will have, in this series, the opportunity of bringing together, from diverse sources, theories and experimental data, and of integrating these into a more general conceptual framework. It is unavoidable, and probably even desirable, that the special bias of the individual authors will become evident in their contributions. Scope will also be given for presentation of new and challenging ideas and hypotheses for which complete evidence is at present lacking. However, the main emphasis will be on fairly complete and objective presentation of the more important and more rapidly advancing aspects of biology. The level will be advanced, directed primarily to the needs of the graduate student and research worker.

Most monographs in this series will be in the range of 200–300 pages, but on occasion a collective work of major importance may be included exceeding this figure. The intent of the publishers is to bring out these books promptly and in fairly quick succession.

It is on the basis of all these various considerations that we welcome the opportunity of supporting the publication of the series '*Frontiers of Biology*' by North-Holland Publishing Company.

E. L. TATUM
A. NEUBERGER, *Editors*

List of contributors

BENNETT, EDWARD L., Laboratory of Chemical Biodynamics, Lawrence Radiation Laboratory, University of California, Berkeley, California, U.S.A.

DOMAGK, GÖTZ F., Department of Biochemistry, University of Louvain, Louvain, Belgium.

DYAL, JAMES A., Department of Psychology, University of Waterloo, Waterloo, Ontario, Canada.

FJERDINGSTAD, EJNAR J., Institute of General Zoology, University of Copenhagen, Copenhagen, Denmark.

GOLUB, ARNOLD M., Walter E. Fernald School, Eunice B. Shriver Kennedy Centre, Waverly, Massachusetts, U.S.A.

HERBLIN, WILLIAM F., Central Research Department, E. I. du Pont de Nemours & Co., Wilmington, Delaware, U.S.A.

JACOBSON, ALLAN L., Department of Psychology, San Francisco State College, San Francisco, California, U.S.A.

KRECH, DAVID, Psychology Department, University of California, Berkeley, California, U.S.A.

MCCONNELL, JAMES V., Mental Health Research Institute, University of Michigan, Ann Arbor, Michigan, U.S.A.

REINIŠ, STANISLAV, Department of Psychology, York University, Toronto, Ontario, Canada.

ROSENTHAL, EUGENE, Department of Pharmacology, University of Minnesota, Minneapolis, Minnesota, U.S.A.

RUCKER, WILLIAM B., Department of Psychology, Mankato State College, Mankato, Minnesota, U.S.A.

SPARBER, SHELDON B., Department of Pharmacology, University of Minnesota, Minneapolis, Minnesota, U.S.A.

UNGAR, GEORGES, Departments of Pharmacology and Anesthesiology, Baylor College of Medicine, Texas Medical Center, Houston, Texas, U.S.A.

WEISS, KENNETH P., Department of Psychology, King's College, Wilkes-Barre, Pennsylvania, U.S.A.

ZIPPEL, HANS P., Physiologisches Institut der Georg August Universität Göttingen, Göttingen, Germany.

Introduction

Learning and memory are phenomena every human being has experienced. Indeed the tremendous ability of our species to learn and to utilize stored information is what most strikingly distinguishes it from even its closest primate relatives. Because of the obvious importance of these processes it might be supposed that a great deal would be known about their biological basis, but until recently very little knowledge had been gained about the processes which must take place inside the central nervous system during and after learning, and which are responsible for the storage of the learned information.

There may be several reasons for this lack of knowledge. No doubt research in the field has to some degree been held back by a more or less conscious acceptance of the (essentially prescientific) attitude that such matters of the 'mind' would not prove open to investigations by ordinary scientific methods. However, the phenomena themselves have also proven surprisingly refractory to seemingly very promising lines of approach. A classic example is Lashley's (1950) report of attempts to investigate the nature and localization of the 'memory trace' by means of a cortical ablation technique. After decades of investigations which failed to yield any positive answers to the questions, he was moved to remark that the results seemed to indicate that 'learning is just not possible'.

Before going into any further discussion it would probably be advantageous to define learning and memory in definite terms. This seemingly easy task has created great difficulty as evidenced by the fact that a standard monograph in the field finds itself using five pages for that purpose (Hilgard and Bower 1966). It has been suggested that since different researchers have often expressed widely divergent opinions of what constitutes learning, we might be better off not using this term in scientific language (Ungar 1970). Still, it may be of advantage to consider the following definition by Miller (1967) of learning in the strict sense: 'Learning is a relatively permanent increase in response strength that is based on previous reinforcement, and that can be made specific to one out of two or more arbitrarily selected stimulus situations'. This definition, because of the emphasis on the role of reinforcement (which may of course be positive or negative, e.g. food reward or shock) rules out types of behavior readily acceptable as learning to those who consider the most important aspect of learning to be the acquisition of new information. A case in point is habituation, in which there is a steadily decreasing tendency to respond on repeated presentation of a stimulus that originally elicited a 100% responding (for instance a startle following a sudden loud noise). Most of the other behavioral procedures described in the present volume do, however, fit into the above definition. Learning may conveniently be divided into two types, classical (Pavlovian) conditioning and operant (Skinnerian) conditioning (e.g. see Hilgard and Bower 1966).

Learning has, in contrast to a surprisingly common anthropocentric belief, been found in nearly all parts of the animal kingdom where serious investigation has been carried out. Protozoa (Applewhite 1968; review by McConnell 1966), coelenterates, platyhelminthes, annelids, molluscs and arthropods, all seem to exhibit learning (see reviews by McConnell 1966 and Eisenstein 1967). Whether or not planarians can learn was long a subject of strong controversy (reviewed by Corning and Riccio 1970; see also the chapter by Jacobson in this volume) that has now been settled in the affirmative. Insects are often considered to be governed completely by instinct, but it has been found that bees (Menzel 1968, 1969; Koltermann 1969) and ants (Rechsteiner 1968) can learn an operant conditioning task and remember surprisingly well. Even in the group of vertebrates there is apparently no simple relation between ability to learn and position on the evolutionary scale such as was maintained earlier (Edinger 1908). Thus fish, even though their brain anatomically is much more primitive, are more similar to mammals and birds in learning ability than are reptiles or amphibians (Rozin 1968).

Although the kind of response that an animal can be trained to perform must depend on the type of animal, since nobody can learn something he is physically unable to do, the training procedures could generally be classified into the categories already mentioned, and there are laws of learning with widespread validity like other biological laws. Human behavior, some think, can also largely be understood (and controlled) within the framework of operant conditioning (Eysenck 1963).

Miller's (1967) definition of learning, cited above, stresses, among other aspects, that the change in behavior must be 'relatively permanent'. This makes it easy to distinguish learning from unrelated phenomena such as fatigue and sensory adaptation that also result in behavioral changes which are, however, only transient. The persistent storage of some type of information which may, under the appropriate conditions, be observed as a behavioral change is what we call memory.

Since the knowledge of the fundamental processes of learning and memory is central to any deeper understanding of higher nervous functions, a large number of partly speculative attempts have been made to formulate a hypothesis for the physiological basis of these phenomena (for a description and discussion of the multitude of proposals see Hilgard and Bower 1966). One line of experimental attack on the problem has been neurophysiological in nature, but in spite of the progress made in this field, there has not been any real breakthrough in the understanding of learning and memory in that area (John 1961; Morell 1961; Laursen 1967) and doubt has been expressed that the problem will ever be solved through such an approach.

On the other hand a good deal of progress has been made in recent years using the approaches of (1) application of agents, chemical or physical, that interrupt or abolish memory, (2) studies of chemical changes in the central nervous system during learning, and (3) experiments on so-called 'chemical transfer of learned information', in which it was found that aspects of trained behavior could be induced in naive animals by injecting brain extracts from trained animals of the same or other species. Approach (3) is the topic of this volume and will be dealt with extensively in the following chapters, for which reason we shall limit ourselves to a few historical comments later in this introduction. Since, however, there is an intimate connection between the transfer concept and the evidence derived from the other approaches, it will be advantageous to consider these studies in a little more detail.

It is well known that memory, once established, is very difficult to abolish. Actually, it seems that in order to do this, one would have to destroy the organism. Maybe this insensitivity to treatments that interfere strongly with other functions is one of the factors that gave rise to the common belief in a 'soul' as something different from and more durable than the rest of the body. However, it has been found that immediately after something has been learned (for a few hours at most), the memory is characteristically much more labile and can be interfered with by a number of different treatments which are ineffective at later stages.

One such treatment is the induction of generalized convulsions, which may be done either by electrical stimulation, so-called electroconvulsive shock, or with chemical compounds that induce convulsions (see review by Jarvik 1970). With both types of treatment it has been found that if animals are trained, and convulsions induced immediately after the end of the training session, or during a relatively short time after, a memory deficit results. If the induction of convulsions is delayed beyond this period, no effect is found. This has been well established in mammals (Thompson and Dean 1955; Pearlman et al. 1967), birds (Lee-Teng and Sherman 1966; Cherkin 1968) and invertebrates, i.e. octopus (Maldonado 1968), and has given rise to the concept of a labile 'short term memory' which is active immediately after training, and which in its turn gives rise to or is followed by permanent or 'long term memory'. There is reason to believe that short term memory consists of a specific type of electrical activity induced in the brain during learning (Sheer 1970); it is this activity which is interrupted by the convulsive treatment. Other types of treatment which interfere with the electrical activity such as anesthesia (Pearlman et al. 1961; Cherkin and Lee-Teng 1965; Dye 1969), hypoxia and hypothermia (Glees 1966) show similar effects, which have been reviewed and discussed by Booth (1967).

Evidence that the information of long term memory might be stored by a chemical, as opposed to an electrical mechanism also come from studies using inhibitors of nucleic acid or protein synthesis. It was found that if RNA synthesis was blocked by administration of 8-azaguanine (Dingman and Sporn 1961) or actinomycin D (Agranoff et al. 1967; Davis and Klinger 1968) memory consolidation was interfered with, although the animals were perfectly able to learn the required problem. Similar results have been obtained using puromycin or acetoxycycloheximide to inhibit protein synthesis (Flexner et al. 1963, 1964, 1967; Agranoff and Klinger 1964; Brink et al. 1966; Barondes and Cohen 1966, 1968; Shashoua 1968; see also review and criticism by Deutsch 1969). Neither of these substances was active if injected after a delay, but had to be given within the same period of time during which convulsive treatment is effective.

The conclusion to be drawn from the inhibitor studies would appear to be that long term memory requires a synthesis of protein (to be preceded, of course, by RNA synthesis), which must take place during the consolidation period. These experiments may be interpreted in two different ways. The one favored by some investigators (Agranoff 1969; Cohen 1970) is that the newly synthesized material is simply needed for growth phenomena like, for instance, the formation of new neuronal connections. The other possible explanation, which is implicit in the transfer concept, is that the new molecules carry, coded into their structure, information essential for the storage of the learned information. In the first case the molecules synthesized might be of a type already present, i.e. the change might be only quantitative; in the second case there would have to be a formation of qualitatively different molecules with composition somehow corresponding to the acquired information.

Evidence that qualitatively different macromolecules are formed during learning comes from studies on the effect of learning on RNA synthesis in the central nervous system. It has been found by a number of investigators, using a variety of approaches, that learning is followed by changes in the base composition of brain RNA. Hydén and his group (Hydén and Egyházi 1962, 1963, 1964; Hydén 1967) analyzed RNA of single neurons and found that learning of different tasks not only increased the amounts of RNA, but that the ratio of adenine to uracil increased significantly. Controls exposed to sensory stimulation without learning showed only the increase in amount of RNA, but no change in composition. Shashoua (1968a, b) found that during learning of a complex motor task in goldfish the uridine:cytidine ratio of brain nuclear RNA increased 96%. Fish that received puromycin showed no such changes, and although they were able to learn the problem, they did not show any retention the following day. Glassman and his co-workers (Zemp et al. 1966; Zemp et al. 1967; Adair et al. 1968a, b; Zemp et al. 1970) have reported that just 15 min of avoidance conditioning (learning to jump to a shelf at the sound of a buzzer to avoid shock) produced as much as a 40% increase in the uptake of labelled uridine into mouse brain polysomes. Appropriate control groups showed this to be a true effect of learning, which could not be produced simply by exposure to shock or buzzer without learning. Even more interesting was the fact that previously trained mice, who already knew the problem, did not show any increase in labelled uridine incorporation when given another 15 min session. Relevant in this connection is also the work of Rapoport and Daginawala (1968), who found that in catfish increased amounts of brain nuclear RNA and various changes of the base composition resulted when they were exposed to olfactory stimulants such as redfish extract, shrimp extract and morpholine, while strong irritants like camphor and menthol changed only amounts of RNA, but not base composition.

The finding that learning leads to the formation of new species of RNA (that then presumably would direct the synthesis of new types of protein) together with the evidence described earlier strongly suggests that memory is stored by means of a chemical mechanism. Such a hypothesis was proposed, with amazing foresight, in the early fifties by Halstead (Katz and Halstead 1950) who was also the first to suggest that memory storage might be analogous to the genetic information storage.

The chemical hypothesis of memory is strongly supported by experimental work indicating the possibility of transferring a learned task from one individual to another by purely chemical means. This was first reported by J. V. McConnell in a paper with the rather startling title 'Memory transfer through cannibalism in planarians' (McConnell 1962). Through a series of studies on learning in these flatworms McConnell and his group had progressed from showing that planaria could learn a conditioned response (Thompson and McConnell 1955), to finding that training was retained through regeneration in both halves of bisected planarians (McConnell et al. 1959), but that it could be abolished by ribonuclease treatment during regeneration (Corning and John 1961).

Under further influence of some of the indirect evidence already cited, McConnell (1962) and coworkers tried to transfer between individual planaria the chemical compound presumably responsible for retention. This experiment met with spectacular success, a highly significant difference being observed; 'cannibals' that had been fed trained 'donors' made more than twice as many responses as controls that were fed untrained or naive donors. Similar studies were later also carried out successfully with extracts more well-defined chemically, and the number of reports of transfer in planaria is now quite large, as may be seen in the chapter by Jacobson in this volume.

The results with planaria aroused a good deal of interest, but even more controversy. Not only were a large number of researchers not ready to accept the possibility of a chemical transfer of learned information, but to many it seemed unlikely that an animal low on the evolutionary scale should be able to learn at all. This controversy now appears to have been settled; Corning and Riccio give an excellent review (1970).

In 1964 Thomas Nissen, Hans Røigaard-Petersen and the present author at the University of Copenhagen became interested in the concept of 'chemical transfer'. To avoid any discussion of whether our experimental subjects were able to learn, we chose to use the standard white rat rather than planarians, hoping that the phenomenon would prove so general as to be found in mammals too. Unknown to us two other groups, those of Stanislav Reiniš at Charles University in Czechoslovakia and Allan L. Jacobson at UCLA were at the same time beginning similar studies, following the same line of thought. Georges Ungar at Baylor College of Medicine in Houston had been led to the transfer concept by a different route, through the study of morphine tolerance (Ungar 1965). Thus it was that in the summer of 1965 four reports of chemical transfer of learned information in mammals appeared nearly simultaneously (Babich et al. 1965; Fjerdingstad et al. 1965; Reinis 1965; Ungar and Oceguera-Navarro 1965). A different line of approach, which might be described as intra-animal transfer was followed at this time by David Albert at McGill University, but only published later (Albert 1966).

From this beginning has evolved the study of the 'transfer' effect in vertebrates that is currently being carried out in a couple of dozen different laboratories, and resulting so far in nearly a hundred reports of the phenomenon.

In the present volume thirteen papers have been collected from researchers who have made significant contributions to the transfer field. All of the four original groups are represented, Allan L. Jacobson has written a review of planarian work, and new approches are described by several authors (i.e. chapters by Götz F. Domagk and Hans P. Zippel, Eugene Rosenthal and Sheldon B. Sparber, and the present author).

From these contributors the readers should be able to arrive at a well informed opinion about what may be called the crucial questions of the transfer field at present: (1) Does the phenomenon really exist, i.e. is the overall evidence statistically convincing? (2) Is it a specific transfer of learned information or some more generalized effect? (3) What is the chemical nature of the active component(s) of the brain extracts?

It is no secret that the field is still a controversial one, and quite a few scientists may yet feel inclined to answer the first question in the negative. In his chapter James Dyal collects all the available evidence, including still unpublished research, and arrives at the conclusion that the evidence overwhelmingly favors the existence of the phenomenon, although he also enumerates a large number of failures to find any effect. An excellent discussion of the validity of the concept is also given in the chapter by William F. Herblin; and Georges Ungar in his chapter stresses the viewpoint of the transfer approach as a bioassay for the memory substrate, beset with all the difficulty and limited reliability of any bioassay, and gives a discussion of the conditions for a successful transfer.

While the majority of the authors in the present volume accept the reality of the transfer phenomenon, we also have a paper by one of the most prominent groups of sceptics, David Krech and Edward L. Bennett, who describe in detail their search for a reliable transfer effect, using a rather unique behavioral procedure. This, the editor thinks, gives a good impression of the tantalizing nature that research in this field often has. Every now and then they obtained significant and even highly significant results, but when an apparently exact replication was carried out, the effect could not be obtained again. The very cautious conclusion drawn by these authors is that their results neither prove nor actually contradict the transfer concept. There is no doubt that we still do not know to a sufficient degree all the experimental parameters that must be brought under control, although some factors have been recognized. The chapters by Arnold M. Golub and James V. McConnell, Stanislav Reiniš, Georges Ungar and the present author all give instances of seemingly trivial aspects of procedures, that have been proven to be crucial.

The second question, whether or not the transfer is specific, and thus truly important for our understanding of learning and memory, is again discussed extensively by Dyal, who describes from a psychological standpoint how specificity should be considered not to be 'either/or', but a question of type and degree of specificity, the strongest possible being specificity within a single stimulus dimension. Another instructive discussion of the different types of specificity is given by Golub and McConnell. In both of these chapters the conclusion is that the crucial questions about specificity have not yet been answered, although there is at least evidence of some degree of specificity such as, e.g. stimulus specificity. Results bearing on this problem are also described by Herblin, Reiniš, William B. Rucker, Ungar, Kenneth P. Weiss and the present author. Rucker discusses the so-called 'reversed effects' sometimes found in experiments on specificity, (see for instance chapters by Herblin and the editor) and concludes that this effect per se cannot be considered to disprove the transfer concept, but may be more comparable to the 'overtraining reversal effect'.

The chemical nature of the active component of the brain extracts has been most intensively studied by Ungar, who has progressed to the point where he is able to describe in his chapter the amino acid sequence of a peptide capable of inducing dark avoidance. Golub and McConnell defend the position that the active component is RNA rather than a peptide. This conflict may be due to the binding of an active peptide to ribosomal RNA during extraction as found by Ungar and the present author and described in Ungar's chapter. Other contributions that throw light on this problem are those of Domagk and Zippel, Reiniš, and Weiss.

Stanislav Reiniš is alone so far in working extensively with the effects of metabolic inhibitors on transfer, and describes a large number of partly unpublished data which supports his derepressor hypothesis of memory transfer.

These thirteen chapters, then, should give the reader an impression of the present state and future promise of the field of 'Chemical transfer of learned information'. The editor hopes that they will succeed in demonstrating that new approaches have been opened to the puzzling problem of information storage in the nervous system. These approaches will certainly be pursued further, and only the future can tell us how far they will take us into that last frontier of biology, higher nervous function.

Acknowledgment

The editor would like to thank all who have helped in the preparation of this volume. Special thanks are due to my coauthors for their readiness to collaborate on the project and for their excellent contributions. Most of the work on the book was carried out when I held a faculty position at the Brain Research Institute and Department of Biochemistry of the University of Tennessee Medical Units in Memphis, Tennessee, U.S.A., and I am much indebted to the director and chairman, Dr. William L. Byrne for providing working facilities and for many helpful discussions and suggestions. I am very grateful to Dr. Larry A. Kepner for reading critically various chapters, and to Mr. Rodney C. Bryant, M.A. for reading the complete manuscript and offering many valuable comments on the behavioral aspects as well as language. I also thank Mr. A. T. G. van der Leij of North-Holland Publishing Company for his help and advice in planning the volume. Finally I would like to express my gratitude to 'Statens Åndssvageforsorg' (Danish Mental Retardation Service) for supporting some of my own research described in this book.

Copenhagen, Denmark E. J. FJERDINGSTAD

References

ADAIR, L. B., J. E. WILSON, J. W. ZEMP and E. GLASSMAN, 1968a, Proc. Natl. Acad. Sci. U.S.A. *61*, 606.

ADAIR, L. B., J. E. WILSON and E. GLASSMAN, 1968b, Proc. Natl. Acad. Sci. U.S.A. *61*, 917.

AGRANOFF, B. W., 1969, Macromolecules and brain function – a 1969 Baedeker. *In*: F. E. Hahn, ed., Progress in molecular and subcellular biology. Springer Verlag, New York, Heidelberg, Berlin, pp. 201–212.

AGRANOFF, B. W., R. E. DAVIS, L. CASOLA and R. LIM, 1967, Science *158*, 1600.

AGRANOFF, B. W. and P. D. KLINGER, 1964, Science *146*, 952.

ALBERT, D., 1966, Neuropsychologia *4*, 49.

APPLEWHITE, P. B., 1968, Nature *219*, 1265.

BABICH, F. R., A. L. JACOBSON, S. BUBASH and A. JACOBSON, 1965, Science *149*, 656.

BARONDES, S. H. and H. D. COHEN, 1966, Science *151*, 594.

BARONDES, S. H. and H. D. COHEN, 1968, Proc. Natl. Acad. Sci. U.S.A. *61*, 923.

BOOTH, D. A., 1967, Psychol. Bull. *68*, 149.

BRINK, J. J., R. E. DAVIS and B. W. AGRANOFF, 1966, J. Neurochem. *13*, 889.

CHERKIN, A., 1968, Fed. Proc. *27*, 437.

CHERKIN, A. and E. LEE-TENG, 1965, Fed. Proc. *24*, 328.

COHEN, H. D., 1970, Learning, memory and metabolic inhibitors. *In*: G. Ungar, ed., Molecular mechanisms in memory and learning. Plenum Press, New York, pp. 59–70.

CORNING, W. C. and E. R. JOHN, 1961, Science *134*, 1363.

CORNING, W. C. and D. RICCIO, 1970, The planarian controversy. *In*: W. L. Byrne, ed., Molecular approaches to learning and memory. Academic Press, New York, pp. 107–149.

DAVIS, R. E. and P. D. KLINGER, 1969, Physiol. Behav. *4*, 269.

DEUTSCH, J. A., 1969, Ann. Rev. Psychol. *20*, 85.

DINGMAN, W. and M. B. SPORN, 1961, J. Psychiat. Res. *1*, 1.

DYE, C. J., 1969, J. Gerontol. *24*, 12.

EDINGER, L., 1908, J. Comp. Neurol. Psychol. *18*, 437.

EISENSTEIN, E. M., 1967, The use of invertebrate systems for studies on the bases of learning and memory. *In*: G. C. Quarton, T. Melnechuk and F. O. Schmitt, eds., The neurosciences. Rockefeller University Press, New York, pp. 653–665.

EYSENCK, H. T., 1963, Brit. J. Psychiat. *109*, 12.

FJERDINGSTAD, E. J., T. NISSEN and H. H. RØIGAARD-PETERSEN, 1965, Scand. J. Psychol. *6*, 1.

FLEXNER, L. B., J. B. FLEXNER and R. B. ROBERTS, 1967, Science *155*, 1377.

FLEXNER, L. B., J. B. FLEXNER, R. B. ROBERTS and G. DE LA HABA, 1964, Proc. Natl. Acad. Sci. U.S.A. *52*, 1165.

FLEXNER, J. B., L. B. FLEXNER and E. STELLAR, 1963, Science *141*, 57.

GLEES, P., 1966, Synapses in the visual pathway, ultrastructure of retinal receptors and of developing neurones (a morphologist's view on the problem of the structural basis of memory). *In*: O. Walaas, ed., Molecular basis of some aspects of mental activity. Academic Press, London, pp. 81–103.

HILGARD, E. R. and G. H. BOWER, 1966, Theories of learning, 3rd. ed. Appleton-Century-Crofts, New York.

HYDÉN, H., 1967, Proc. Amer. Philos. Soc. *111*, 326.

HYDÉN, H. and E. EGYHÁZI, 1962, Proc. Natl. Acad. Sci. U.S.A. *48*, 1366.

HYDÉN, H. and E. EGYHÁZI, 1963, Proc. Natl. Acad. Sci. U.S.A. *49*, 618.

HYDÉN, H. and E. EGYHÁZI, 1964, Proc. Natl. Acad. Sci. U.S.A. *52*, 1030.

JARVIK, M. E., 1970, The role of consolidation in memory. *In*: W. L. Byrne, ed., Molecular approaches to learning and memory. Academic Press, New York, pp. 15–26.

JOHN, E. R., 1961, Ann. Rev. Physiol. *23*, 451.

KATZ, J. J. and W. C. HALSTEAD, 1950, Comp. Psychol. Monogr. *20*, 1.

KOLTERMANN, R., 1969, Z. Vergl. Physiol. *63*, 310.

LASHLEY, K. S., 1950, Sympos. Soc. Exp. Biol. *4*, 454.

LAURSEN, A. M., 1967, Ann. Rev. Physiol *29*, 543.

LEE-TENG, E. and S. M. SHERMAN, 1966, Proc. Natl. Acad. Sci. U.S.A. *56*, 926.

MALDONADO, H., 1968, Z. Vergl. Physiol. *59*, 25.

MCCONNELL, J. V., 1962, J. Neuropsychiat. *3* (suppl. 1), 42.

MCCONNELL, J. V., 1966, Ann. Rev. Physiol. *28*, 107.

MCCONNELL, J. V., A. L. JACOBSON and D. P. KIMBLE, 1959, J. Comp. Physiol. Psychol. *52*, 1.

MENZEL, R., 1968, Z. Vergl. Physiol. *60*, 82.

MENZEL, R., 1969, Z. Vergl. Physiol. *63*, 290.

MILLER, N. E., 1967, Certain facts of learning relevant for the search for its physical basis. *In*: G. C. Quarton, T. Melnechuk and F. O. Schmitt, eds., The neurosciences. Rockefeller University Press, New York, pp. 643–652.

MORELL, F., 1961, Physiol. Rev. *41*, 443.

PEARLMAN, C. A., S. N. SHARPLESS and M. E. JARVIK, 1961, J. Comp. Physiol. Psychol. *54*, 109.

RAPOPORT, D. A. and H. F. DAGINAWALA, 1968, J. Neurochem. *15*, 991.

RECHSTEINER, A., 1968, C.R. Acad. Sci. Paris *267*, 1535 (ser. D).

REINIŠ, S., 1965, Actv. Nerv. Sup. *7*, 167.

ROZIN, P., 1968, The use of poikilothermy in the analysis of behavior. *In*: D. Ingle, ed., The central nervous system and fish behavior. University of Chicago Press, pp. 181–192.

SHASHOUA, V. E., 1968a, Nature *217*, 238.

SHASHOUA, V. E., 1968b, The relation of RNA metabolism in the brain to learning in the goldfish. *In*: D. Ingle, ed., The central nervous system and fish behavior. University of Chicago Press, pp. 203–213.

SHEER, D. E., 1970, Electrophysiological correlates of memory consolidation. *In*: G. Ungar, ed., Molecular mechanisms in memory and learning. Plenum Press, New York, pp. 177–211.

THOMPSON, R. and W. DEAN, 1955, J. Comp. Physiol. Psychol. *48*, 483.

THOMPSON, R. and J. V. MCCONNELL, 1955, J. Comp. Physiol. Psychol. *48*, 65.

UNGAR, G., 1965, Fed. Proc. *25*, 548.

UNGAR, G., 1970, Molecular mechanisms in memory and learning. Plenum Press, New York, pp. v–x.

UNGAR, G. and C. OCEGUERA-NAVARRO, 1965, Nature *207*, 301.

ZEMP, J. W., J. E. WILSON and E. GLASSMAN, 1967, Proc. Natl. Acad. Sci. U.S.A. *58*, 1120.

ZEMP, J. W., J. E. WILSON, K. SCHLESINGER, W. O. BOGGAN and E. GLASSMAN, 1966, Proc. Natl. Acad. Sci. U.S.A. *55*, 1423.

Contents

CHAPTER 1

Empirical issues in interanimal transfer of information

ARNOLD M. GOLUB* and JAMES V. McCONNELL

Mental Health Research Institute, University of Michigan, Ann Arbor, Michigan

1.1. Introduction

What are the mechanisms underlying learning and memory? Although a large number of hypotheses have been suggested, few of these have been stated precisely enough to be testable in the laboratory. Hypotheses about the physical basis of memory generally fall into one of two categories, those in which macromolecules are posited as information coding mechanisms and those in which the importance of neural connections is stressed.

In recent years, considerable evidence has accumulated that suggests that long-term memory is not coded in terms of electrical activity. In the face of this evidence, a number of scientists have concluded that the long-term retention and retrieval of experiential information is probably mediated by permanent chemical changes in the central nervous system (Katz and Halstead 1950; Barondes 1965; Booth 1967). Nevertheless, the bulk of the empirical research concerned with elucidating the physical basis of memory suggests that the chemical code, if it exists, is highly elusive.

In fact, almost all available evidence for a chemical 'memory' code is largely indirect. Studies in which electroconvulsive shock or other disrupting agents are used to study memory have provided the best evidence against the proposition that long-term information is held in perseverating electrical activity. Yet, such studies have done little to provide direct evidence for a molecular storage mechanism in the long-term retention and retrieval of experiential information.

Similarly, studies of changes in ribonucleic acid (RNA) base ratios during learning (Hydén 1967), although providing some evidence for a molecular storage mechanism for information, require extremely sophisticated pro-

* Present adress: Walter E. Fernald School, Eunice B. Shriver Kennedy Center, Waverly, Massachusetts.

cedures that have prevented independent replication of the work in other laboratories. Even if these very elegant techniques were generally available, the task of differentiating RNA base ratio changes that are correlated with learning from those changes occurring as the result of non-specific factors, such as activity or stress, still needs to be undertaken. Furthermore, these base ratio changes apparently are not permanent, so investigators using this approach have the additional problem of making inferences from transient changes in RNA base ratios to the complexity of unknown steps that must occur following these ephemeral changes.

Recently, three different approaches to elucidating the chemical correlates of memory have been undertaken, and these show promise of more directly implicating any contribution macromolecules might make to the physical storage of information.

The first of these approaches was initiated by McConnell and his colleagues at the University of Michigan, and was prompted by the unexpected observation that when trained flatworms were transversely sectioned and allowed to regenerate, both of the newly formed worms showed savings on the task in which the original worm had been trained. This finding was of particular interest, since only one of the newly formed worms contained the brain of the original. Later experiments indicated that cannibal worms fed on trained worms and subsequently trained themselves, acquired a conditioned response significantly faster than did cannibals ingesting untrained material (McConnell et al. 1961).

Subsequent research with the planarian (Corning and John 1961) indicated that the substance responsible for the 'second generation' transmission of acquired information is probably RNA, and that the effect can be eliminated if RNA extracted from the trained donor worms is incubated with RNAse prior to injection into recipient worms (McConnell 1968).

The second approach to studying the biochemistry of memory was begun in the laboratory of Edward Glassman at the University of North Carolina by Zemp and his colleagues (1966, 1967). Through the use of radioactive labeling procedures, Zemp and his associates demonstrated increased incorporation of labeled precursors of RNA during a learning experience in mice. The initial studies reported by these investigators did not include control groups necessary for the conclusion that the observed changes in synthesis of RNA during acquisition were specific to learning experience and not to non-specific factors. More recent radioactive pulse-labeling data from that laboratory (Adair et al. 1968), however, has provided some strong evidence that the increased synthesis of RNA during learning is probably due to the learning experience.

The third approach to more directly investigating biochemical processes in memory has been used exclusively by Gaito and his colleagues at York University (Machlus et al. 1968, 1969). In this approach DNA–RNA successive competition hybridization procedures are applied to learning situations to determine whether unique species of RNA are synthesized during learning. Since the basic hybridization procedure was developed with bacteria (Gillespie and Spiegelman 1965), it is too early to determine whether its application to mammalian brain tissue as a memory assay is entirely appropriate. However, if the work of Gaito can be repeated by other investigators, his ingenious application of the hybridization procedure to assay novel RNA synthesized during learning may prove a highly significant tool for elucidating macromolecular changes during learning.

The 'memory transfer', pulse-labeling and successive competition hybridization approaches to defining the biochemical substrate of long-term information storage have one factor in common. They all appear to be reasonably direct methods for studying changes in brain chemistry during learning and for referring these changes to particular substances whose modification is directly correlated with the learning experience. With the memory transfer approach, a series of incompletely known steps must still be considered (e.g. uptake, incorporation and metabolic fate of the injected substance); nevertheless, the transfer approach does lend itself to direct coordination of observed changes in recipient behavior with the substance contained in the injection. Similarly, by making use of pulse-labeling or hybridization procedures, one can coordinate changes in behavior that occur as a function of the learning experience with changes in the synthesis of RNA. Thus, investigators using any of these three approaches, in addition to having sophisticated and sensitive techniques that are nevertheless available to other workers interested in repeating or extending the experiments, have a direct method for referring biochemical changes occurring during learning to memory mechanisms.

In 1965, at least four independent research groups applied the planarian memory transfer paradigm to mammals and reported a striking degree of success in transferring experiential information to naive recipient mice or rats via injections of material extracted from the brains of trained animals

(Babich et al. 1965; Fjerdingstad et al. 1965; Reiniš 1965; Ungar and Oceguera-Navarro 1965). These initial reports of memory transfer were followed by a series of negative results from other laboratories (Gross and Carey 1965; Byrne et al. 1966; Luttges et al. 1966). Although additional positive results were reported, these also were soon followed by papers in which investigators described their failures to repeat the basic observation (see Jacobson 1967; Schutjer 1968; and Tunkl 1968 for annotated bibliographies of transfer studies).

The mammalian memory transfer studies have now been repeated some 127 times in at least 30 laboratories and have resulted in even more controversy than that generated by the planarian experiments. This controversy appears to be directed towards three questions: Is the transfer effect a reliable, repeatable phenomenon? How specific is the effect? And what is the active substance(s) mediating the effect? In this chapter, we will discuss each of these questions in turn and report a series of experiments carried out in our laboratory at the University of Michigan that have led us to the conclusion that the transfer phenomenon does represent a valid experimentally reproducible effect that may prove of considerable value in better defining the contribution of macromolecular mechanisms in memory.

1.2. Reliability

In 1967, a group of investigators met in Chicago to hold a work session on memory transfer. The primary purpose of this session was to develop and check paradigms for memory transfer in mammals that could be repeated by independent investigators in their own laboratories. A number of the participants agreed to test one or more of the paradigms presented at this meeting and to report their results at a later date. Subsequently, the results of these attempts were presented at the American Association for the Advancement of Science meetings in 1967 in New York.

With some notable exceptions, it appeared that even in laboratories where positive results had previously been obtained, investigators failed to find evidence for the transfer phenomenon when they repeated the work of another investigator. The issue became even more confusing when it was reported that memory transfer experiments performed with material extracted from the brains of animals trained in one laboratory yielded positive transfer effects in a second laboratory, but that investigators in that second laboratory failed to find reliable transfer effects when they themselves trained the donor animals.

As part of the AAAS program, we agreed to repeat the experiments of Gay and Raphelson (1967) as modified and used extensively by Ungar and his associates at Baylor College of Medicine (Ungar et al. 1968). In this type of experiment, donor animals are trained to escape from a black into a white chamber to avoid a painful electric shock. When extracts from the donor brains are injected into untrained recipients, and the recipients placed for the first time into the apparatus, the rats that are injected with substances from trained donor brains typically show a statistically reliable tendency to avoid the black chamber and to approach the white chamber. Control recipient animals, injected with material from untrained donor brains, show the 'natural' reaction of avoiding the white box and remaining in the black chamber.

Our initial replication of the foregoing paradigm was highly successful; however, following this initial success, we were unable to find differences between experimental and control groups of recipients in the next five experiments.

1.2.1. Incubation effects in memory transfer

After failing to find an acceptable interpretation for these failures, we turned to a paradigm that was first reported by Dyal and his associates (1967) and replicated successfully in our laboratory (Golub and McConnell 1968). In these experiments training of the donor takes place in an operant chamber, in which the animal is trained to press a bar in order to obtain food reward. Following eight to ten days of bar-press training on a continuous reinforcement schedule, the donor animals are given three days of experimental extinction during which bar presses are not reinforced with food, followed in turn by three days of reacquisition training in which bar presses again deliver food. A second group of donor animals serves as a control group for handling and sensitization. Following training sessions, donor animals are sacrificed, and homogenates prepared from their brains are injected i.p. into untrained recipient rats. Memory transfer effects are consistently reported for animals that receive material from trained animals.

Late in 1968, we began what was designed to be little more than a replication of this prior work by Dyal and his associates. However, the results of our first experiment were so unanticipated that we subsequently undertook several additional studies to confirm our findings (Golub et al. 1970).

We placed naive, male, Sprague–Dawley rats on a food deprivation schedule for six days and then trained them to bar press for food in an operant chamber that had been modified so that the bar was moved to the side of the chamber opposite the food cup and centered about 1 in from the door of the chamber. Following eight days of bar-press training, in which each bar press delivered a 45 mg Noyes pellet into the food cup, we administered three days of extinction in which bar presses no longer produced food, followed by three days of reacquistion in which bar presses again produced food. During all these training sessions, the animals were trained for 30 min and received one session each day. We assigned other animals to a donor control group. These control donor animals were handled each day and maintained on the same food schedule as were the experimental donor group, but they were not trained.

Within 15 min following its final training session, each of the experimental donors was sacrificed by decapitation. The brain, excluding the olfactory bulbs and cerebellum, was removed and stored on dry ice. The control donors were similarly sacrificed. These brains were assigned code numbers, then were homogenized by adding 1 cc of 0.154 M NaCl per brain and homogenizing gently for 15 strokes using a motor driven pestle, with an ice bath surrounding the tissue grinder.

Recipient rats that had been placed on 22.5 hrs food deprivation schedules six days earlier were lightly etherized and injected i.p. with 3.2 cc of one of the coded homogenates. These recipient animals received their first test for transfer 24 hrs after injection. Testing was carried out using the procedures developed by Dyal et al. (1967). During the initial 30 min, the bar was removed from the operant chamber and the food dispenser activated every 60 sec. The number of nose entries into the food cup was automatically recorded using a photocell installed into the food cup. At the end of this 30 min test, the animal was removed from the chamber, the bar was reinserted, and the rat was given a second 30 min test in which each bar press was reinforced with food. On each of the following eight (continuously reinforced) days, these recipient animals received daily 30 min bar-press sessions. The mean number of bar presses on each of the nine days of bar-press training is presented in fig. 1.1. As can be observed from fig. 1.1, the two curves begin to separate after day 2.

On days 8 and 9 of training the mean number of bar presses was 106.8 and

129.8 for animals receiving brain material from experimental donors. The mean number of bar presses on days 8 and 9 was 216.2 and 187 for animals receiving brain material from control donors. The differences between the groups on both of these days are statistically reliable ($p=0.02$) with the Mann–Whitney U test. Thus, the control animals (injected with homogenate from untrained brains) were reliably superior to the experimental recipients

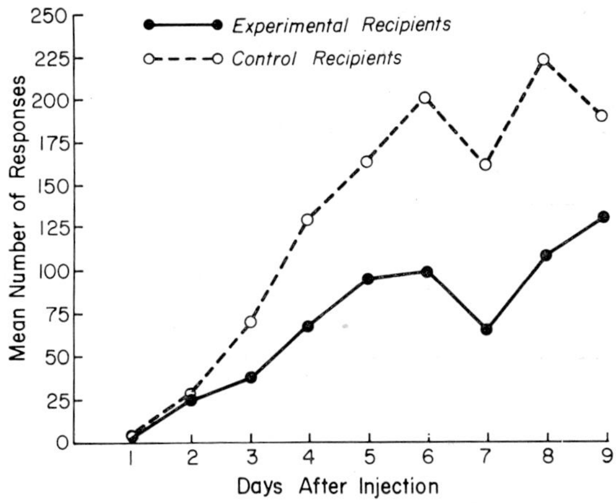

Fig. 1.1. Mean number of bar presses on each of nine daily sessions for recipient animals injected either with brain homogenate from the experimental or from the control donor groups. Each point is based on six subjects (60- to 90-day-old male Sprague–Dawley rats).

(injected with homogenate from trained brains), as measured by bar pressing. This result was unexpected, since in our previous replication of this paradigm (Golub and McConnell 1968), we found that experimentally injected recipients were superior to control injected recipient animals on the first (and only) postinjection test.

There was however, one major procedural difference between the present study and our 1968 experiment. In the present study, due to difficulty in obtaining recipient animals from the supplier, training of the donor animals was discontinued for one week following the three days of extinction training, so that recipient animals would be available at the exact time that the donors were sacrificed. During this seven-day period, the experimental donor animals were handled, but were not trained. It occurred to us that this seven-day rest period, during which the animals received no training, might have functioned as an 'incubation' period for the training experience just preceding it (i.e. the extinction training), so that the learned tendency 'not to press the lever' was perhaps the most likely aspect of the training to transfer. We decided to test this notion by repeating the experiment a second time and by adding to the design a third donor group, in which we interposed a rest period at a different point in the donor training regime. The revised paradigm for this experiment is presented in table 1.1.

TABLE 1.1

Paradigm for memory transfer – incubation experiments.

Donor groups	Treatment
Experimental group 1	Acq (8 days) + REST (7 days) + Ext (3 days) + Reacq (3 days)
Experimental group 2	Acq (8 days) + Ext (3 days) + REST (7 days) + Reacq (3 days)
Handling control	Handled each day, but not trained

One group of donor rats received eight daily sessions of acquisition, followed by a seven-day rest period during which it was not trained, followed by three daily sessions of extinction in which bar presses did not produce food, followed by three daily sessions of reacquisition (Acq-Rest-Ext-Reacq). A second group of donor animals received the same training as the foregoing group, except that a rest period of seven days was not interposed into the training regime until extinction training had been completed (Acq-Ext-Rest-Reacq). A third group of donors served as controls for handling and was not trained. Following training, all donor animals were sacrificed, and homogenates were prepared in the manner described in experiment 1. Three groups of recipient rats were lightly etherized and injected with 3.2 cc of homogenate per recipient animal. Brain extraction, homogenization and injection of recipient animals were carried out using 'blind' procedures to avoid introducing experimenter bias.

Testing of recipients was performed in the manner described in experiment 1. This second experiment was repeated an additional time and the results of the two replications are presented in figs. 1.2 and 1.3.

As can be seen from figs. 1.2 and 1.3, recipient animals injected with homogenate from donor animals receiving a rest period following acquisition

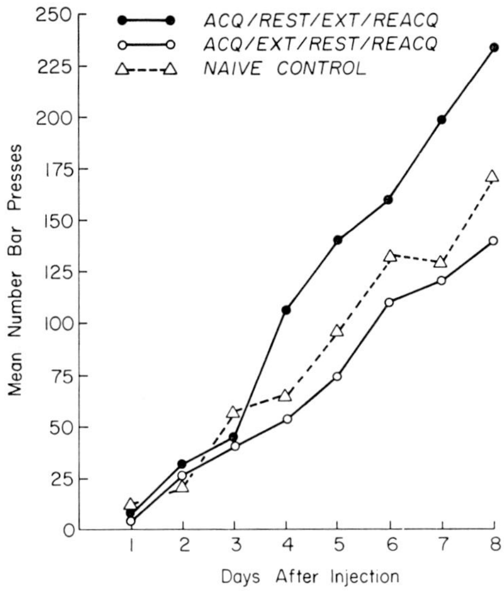

Fig. 1.2. Mean number of bar presses emitted by the various recipient groups on each of eight training days following injection of homogenate prepared from the brains of either the experimental or control donor animals. Each point is based on six subjects (60- to 90-day-old male Sprague–Dawley rats).

training, but prior to extinction (Acq-Rest-Ext-Reacq) appear to acquire the bar-press response more rapidly than do recipient animals in either of the two other groups, and recipient animals injected with material from donors that had received a rest period following extinction and prior to reacquisition (Acq-Ext-Rest-Reacq) appear to be inferior to the other groups in terms of rate at which they acquire the bar-press response. The differences between the groups on these studies do not quite reach statistical significance at the conventional level. At this point, we decided against simply repeating the experiment an additional time in order to achieve statistical reliability by increasing the sample size. Although we were convinced that we were dealing with a real phenomenon, we wanted to increase the size of the effect so that statistically reliable results would be found with small sample sizes. Since in the past we had often obtained what appeared to be much more reliable effects using RNA-rich extracts rather than whole brain homogenates, we decided to repeat the paradigm an additional time, on this occasion extracting an

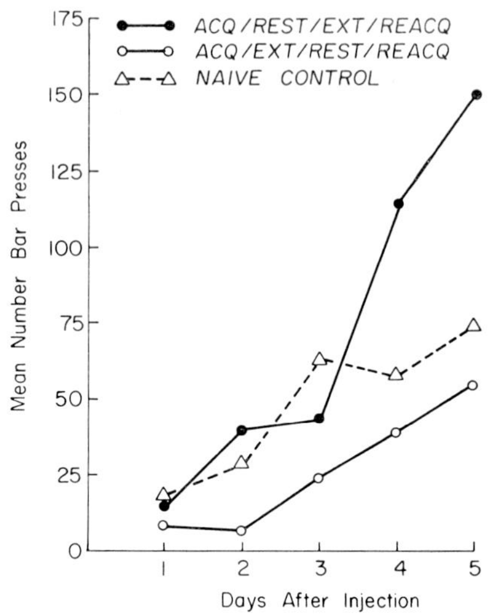

Fig. 1.3. Mean number of bar presses emitted by the various recipient groups on the five training days following injection of brain homogenate from either experimental or control donor animals. Each point is based on six subjects (90- to 110-day-old male Sprague–Dawley rats).

RNA-rich fraction from the donor brains and injecting it directly into the brains of the recipients. We felt that in this way we could increase the size of the effect so that conventional statistical procedures would yield reliable effects (even with very small sample sizes).

We therefore again trained three groups of donor animals (table 1.1) using the same procedures mentioned above. We hypothesized that if rest periods did indeed serve an incubatory function of consolidating the learning experience just preceding them, recipient animals receiving brain material from donors given their incubation period following acquisition training (Acq-Rest-Ext-Reacq) should be superior in their rate of learning the bar-press response to recipient animals injected with material from donors in group Acq-Ext-Rest-Reacq.

Following the final day of donor training, all donor animals were sacrificed by decapitation and an RNA-rich extract prepared from their brains. Sacrifice of the donor animals, RNA extraction, and injection and testing of the recipient animals were all performed using 'blind' procedures.

RNA was obtained by a cold phenol extraction procedure; all steps were carried out at 0–4 °C unless otherwise stated. For each gram of tissue, 10.0 ml of 0.154 M NaCl were added. The brains were allowed to thaw partially and were then homogenized for 1.0 min using a motor-driven Teflon pestle. An ice bath jacketed the homogenizer during this process. Following homogenization, 0.5 volume of 88% phenol was added and the homogenate stirred with a magnetic stirring bar for 30 min. The resulting mixture was centrifuged at 20,000 *g* for 60 min. The top, aqueous phase was carefully withdrawn and transferred to clean test tubes. Then 0.10 volume of 1.0 M $MgCl_2$ was added and the solution stirred gently. Following this, 2.50 volumes of cold 95% ethanol were added and the solution stirred. This mixture was stored at −20 °C for 2.0 hrs. The resultant precipitate was sedimented at 1400 *g* for 20 min. The supernatant was decanted and the precipitate resuspended in 6.0 ml of 75% ethanol and resedimented by centrifugation at 1400 *g* for 15 min. Resuspension and sedimentation were repeated three times. The product was dried by evaporating residual ethanol in an air stream and was then dissolved in 0.60 ml of 0.154 M NaCl per brain equivalent of extract.

The product derived from this isolation procedure has been evaluated by several methods (Seifter et al. 1950; Lowry et al. 1951; Ashwell 1957; Ames et al. 1960). The extract contains RNA, negligible amounts of DNA and protein and polysaccharide contaminants. A complete quantitative analysis of the extract awaits further study. The yield of product from one brain reserved for assay was dissolved in 10.0 ml of gradient buffer containing 0.01 M sodium acetate, pH 5.1, 10^{-3} M Na EDTA, and 0.1 M NaCl. An ultraviolet absorption spectrum of this solution is presented in fig. 1.4. Two

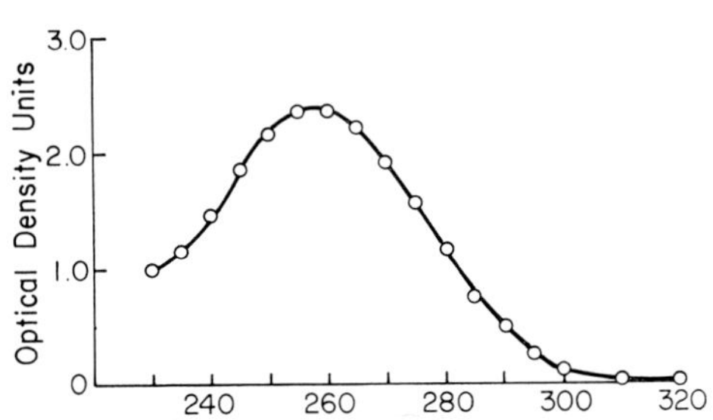

Fig. 1.4. Ultraviolet absorption spectrum of one brain equivalent of RNA-rich extract dissolved in 10.0 ml of 0.01 M acetate buffer, pH 5.1 with 10^{-3} M EDTA and 0.1 M NaCl.

ml of the solution was layered on a 28 ml linear gradient of 10–40% RNAse-free sucrose in the gradient buffer and centrifuged at 25,000 rpm for 15 hrs at 0 °C in an SW 25.1 rotor of a Spinco L2-65B ultracentrifuge. The gradient was analyzed for the distribution of 260 mμ absorbing material using an LKB quartz flow cell with a 4 mm light path modified for use in a Beckman DU spectrophotometer with Gilford components and continuous recording on a Sargent recorder.

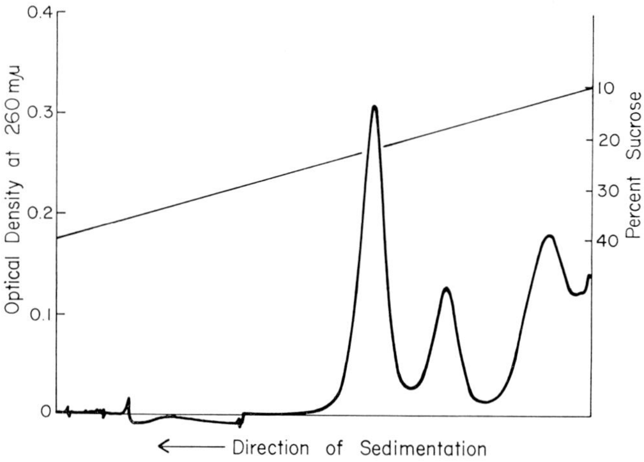

Fig. 1.5. 10–40% linear sucrose density gradient sedimentation profile of $\frac{1}{5}$ brain equivalent of RNA-rich extract. The gradient contained 0.01 M acetate buffer, pH 5.1, with 10^{-3} M EDTA, and 0.1 M NaCl and was centrifuged in a Spinco L2-65B ultracentrifuge in an SW 25.1 rotor.

The density gradient optical density profile (fig. 1.5) indicated the presence of both 28S and 18S ribosomal RNA and transfer RNA. The ribosomal RNA appeared to be undegraded, as indicated by the 28S:18S peak area ratio which was greater than 2.0. No tests of the integrity of the other components of the extract were performed.

Recipient rats were anesthetized with sodium pentobarbitol. An incision was made along the line of the sutura sagittalis and the skull was exposed. A dental drill with a 1 mm burr was used to make a hole in the sutura sagittalis between the eyeballs and above the olfactory bulbs. A 0.25 ml syringe was used to inject the RNA-rich extract. This syringe contained 0.15 ml (0.25

brain equivalents) of extract. The needle was kept at a 45° angle with the frontal plane and 2 mm deep from the outside surface of the skull. The extract was injected over a 5 min period.

Twenty-four hrs following these subdural injections, the recipient animals were given the first of eight daily (30 min) bar-press sessions, during which each bar press was reinforced with a Noyes pellet. The mean number of bar presses emitted by each of the groups during each of these eight sessions is presented in fig. 1.6. An unweighted-means solution analysis of variance

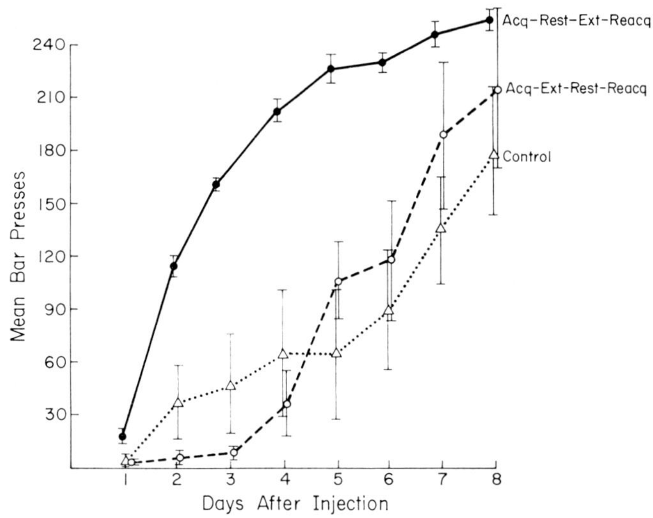

Fig. 1.6. Mean number of responses emitted by the various recipient groups on each of eight days following injection of RNA-rich brain extract. The RNA was extracted from the brains of donor animals trained on either Acq-Rest-Ext-Reacq, Acq-Ext-Rest-Reacq, or from untrained control animals. The vertical bars represent one standard error of the mean. Each point is based on four subjects (60- to 90-day-old male Sprague–Dawley rats).

was used to evaluate the group means during the initial three days of training. The analysis indicated that the three groups were reliably different in their rate of bar pressing ($p<0.05$) and that there was a significant interaction between sessions and treatments ($p<0.05$). Individual comparisons between the three groups indicated that recipients injected with material from Acq-Rest-Ext-Reacq donors were reliably superior to either of the other recipient groups ($p<0.05$) and that for the first three days of training recipients injected with material from untrained donors were reliably superior to animals injected with extract from Acq-Ext-Rest-Reacq donors in mean number of responses.

These results suggest that incubation periods may play more than a casual role in determining whether transfer effects occur with certain paradigms. Reiniš (1969) has informed us that he obtains transfer effects with his paradigms when he trains his donor animals six days a week but not when they are given additional training sessions on the seventh day. It would appear possible that this day of no training may function to incubate the experimental material and that a proportion of the negative transfer studies in the literature reflects the failure of certain investigators to incorporate incubation periods into the training regimes of their donor animals. This is not to say that incubation periods interposed during donor training are critical for memory transfer to occur with all paradigms presently extant – it is clear that they are not. There is abundant evidence in the literature indicating that if additional treatments, such as interpolation of extinction trials into acquisition paradigms or significant training of donor animals beyond criterion levels of performance (overlearning), are incorporated into transfer paradigms, these incubation periods are unnecessary for transfer effects to occur. In the absence of the foregoing, or similar, treatments it seems reasonable to infer that incubation periods do play a role in determining whether transfer effects are found.

In summary, our data suggest that incubation periods may be of considerable importance in memory transfer studies. In any area in which important variables have yet to be defined adequately, early experimentation typically yields effects smaller than those found in later work when more is known about such variables. Were this the situation with the incubation phenomenon, it would account for the lack of repeatability of some paradigms by conscientious scientists who, in attempting to repeat the original transfer experiments, train their donor animals on the basis of a seven-day week, unaware that some of the best evidence for the phenomenon has been obtained when donor animals are trained only five or six days a week (Byrne et al. 1966; Hoffman et al. 1967; Corson and Enesco 1968; Hutt and Elliott 1970).

1.2.2. Incubation effects in learning

We do not know why these incubation periods should be effective in modulating transfer effects. One possibility, however, is that incubation functions by increasing the strength of learning in donor animals, and therefore increasing the probability that the transfer effect will be found in recipient animals. Further, it would appear important as a first step in investigating the specificity of the transfer phenomenon, to determine whether variables that affect donor learning also affect memory transfer. For these reasons we decided to carry out a series of experiments in which incubation periods were incorporated into learning paradigms. In these experiments 'strength of learning' was measured either by resistance to extinction or by the number of responses emitted during an acquisition session of specified length.

We decided first to investigate the effect on subsequent resistance to extinction of incubation periods interposed during acquisition training. Sprague–Dawley rats were all given an initial session of food-magazine training in our operant chambers (described above), followed by four sessions of acquisition training on a CRF schedule. At the end of day 4, all subjects were systematically assigned to one of four groups on the basis of their bar-press rates. Animals in group 1 received five days rest, during which they were given no training, followed by two additional days of acquisition training (Acq 4-R5-Acq 2); animals in group 2 received additional training on each of the following seven consecutive days (Acq 11); animals in group 3 received an additional two days of training, followed by a five-day rest period (Acq 6-R5); animals in the fourth group received two additional days of training, but no rest period (Acq 6). On the day following the last training session, or rest day, depending upon to which group the animal had been assigned, all subjects received four 30 min sessions, one session a day, of experimental extinction. During these extinction sessions bar presses were not reinforced with food.

The results of this experiment are presented in fig. 1.7.

We found no statistically reliable differences between any of the groups on the bar-press measure on the final day of extinction. However, the extinction data provided evidence that the groups were not equivalent at the end of acquisition. The mean number of bar presses made on the first day of extinction was 140, 134, 110 and 106 for animals trained on Acq 4-R5-Acq 2, Acq 11, Acq 6-R5, and Acq 6, respectively. The difference between group Acq 4-R5-Acq 2 and group Acq 6 was statistically reliable ($t=2.2$, $p=0.05$). No differences were found between any of the other groups on the first day of extinction. On day 2 of extinction the difference between the mean number of bar presses for group Acq 4-R5-Acq 2 and group Acq 6 was statistically reliable ($t=3.03$, $p<0.05$). Again, no differences were found between any of the other groups. On day 3 of extinction, there were no reliable differences between groups Acq 4-R5-Acq 2, Acq 11, and Acq 6-R5; however, both the Acq 11 and the Acq 4-R5-Acq 2 groups were reliably different from the Acq 6 group ($t=2.52$, $p<0.05$ and $t=2.58$, $p<0.05$, respectively).

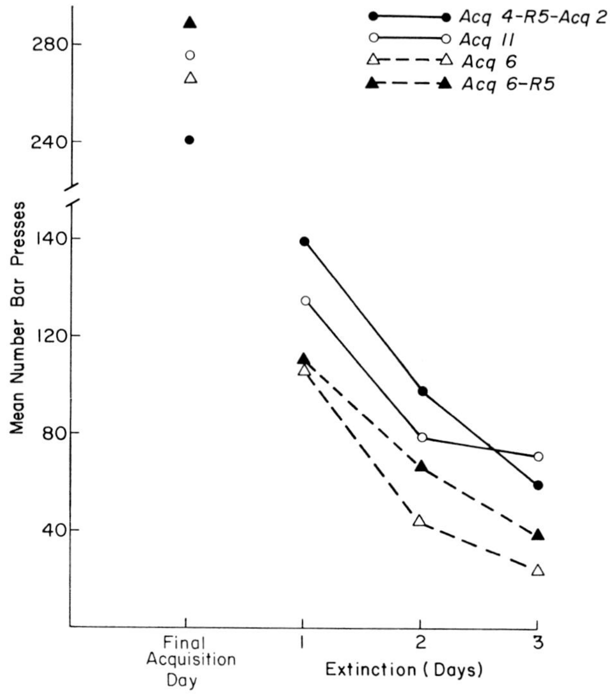

Fig. 1.7. Mean number of bar presses emitted by the various groups during the final acquisition session and on each of the extinction sessions. Each point represents the mean of seven subjects (60- to 90-day-old male albino rats).

In a second experiment we studied the role of incubation in acquisition of the bar-press response by making use of standard food-magazine training procedures. As we mentioned earlier, the levers in our operant chambers had been moved to the wall opposite the food cup. Under these conditions, it is difficult for a rat to learn to bar press, since the rat may take a great deal of time to find the food after it initially presses the bar and may never learn to associate bar pressing with food delivery. In all our previous operant conditioning studies, we typically gave our experimental donor animals an initial session of food-cup (food-magazine) training during which the food dispenser was automatically activated once each min independently of the rat's behavior, and food was delivered into the food cup. In this way the animal learned to associate the click of the food dispenser with food delivery. Following this initial food-magazine training session, when the rat was placed into the apparatus and in the course of its exploration hit the bar, the distinct click of the food dispenser directed the rat to the food cup and the animal acquired the bar-press response readily.

Earlier experiments in our laboratory had indicated that the male albino rat needs at least 20 click–food pairings to learn the relationship between the click and food. We made use of this fact by carrying out the following experiment. We assigned two groups of rats to either an incubation or a non-incubation condition. Half the animals in each of these groups received a 20 min food-magazine training session. The other half received a 10 min food-magazine training session. During this session, a 45 mg Noyes pellet was delivered into the food cup once each min. Following this single magazine training session, all animals were returned to their home cages. Animals in the incubation group rested for two days following their magazine training experience. After this two-day rest period, they were given a single 30 min bar-pressing session for food on a continuous reinforcement schedule. The animals assigned to the non-incubation group were given a single 30 min bar-pressing session on the following day during which each bar press was rewarded with food.

We hypothesized that if incubation were actually affecting learning, we should see no effect of incubation when animals received a 10 min magazine-training session, since our earlier work had shown that in the 10 min period rats learn very little that they could subsequently incubate. On the other hand, we hypothesized that animals given a 20 min magazine-training session followed by incubation, should acquire the bar-press response significantly faster than animals receiving the same amount of magazine training, but no incubation.

The results are presented in fig. 1.8. As can be seen from fig. 1.8, there are no differences between the groups on the bar-press measure when a 10 min magazine-training session is used; however, when the length of the magazine-training session is increased to 20 min, the group that had received the two-day incubation period is clearly superior to the non-incubation group ($t=2.1$, $p<0.05$).

The results of the two foregoing studies provide evidence for the importance of incubation as a variable that facilitates learning. When the incubation phenomenon is considered as it relates to memory transfer, these data suggest

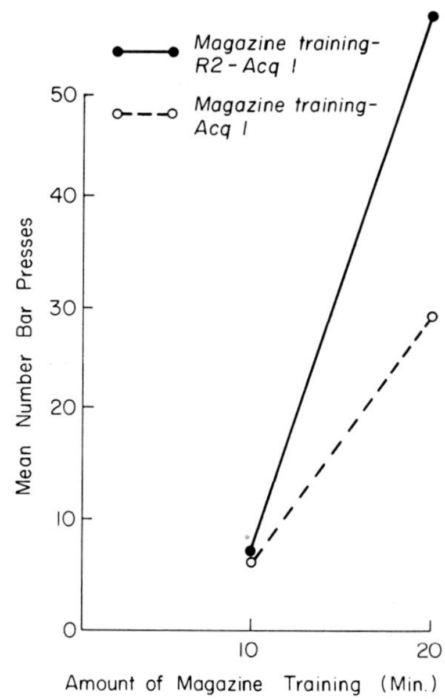

Fig. 1.8. The effect of two intervals of food-magazine training on bar-pressing following either no incubation or a two-day incubation period (no incubation is defined as any intersession-interval no longer than 24 hrs). Each point is based on 14 subjects (60- to 90-day-old male Sprague–Dawley albino rats).

that it is possible to coordinate factors that are important in donor training with the type and extent of memory transfer effects one obtains. As we have mentioned earlier, if the transfer phenomenon does involve transfer of specific information from donor to recipient animals, then it should (must?) be the case that variables affecting donor learning will also affect the transfer phenomenon. The experiments presented in this chapter suggest that a coordination between donor training and recipient transfer effects is possible and also suggests that it should be possible to obtain transfer effects with some paradigms used unsuccessfully in the past, provided that one includes incubation periods in the donor animal training paradigm.

1.3. Specificity of the transfer effect

The most extreme statement of a macromolecular theory of memory is posed by the memory transfer experiments. These experiments, and the logic underlying them, are probably the least understood and most misinterpreted work in the experimental literature. The purpose of this section is to attempt to clarify the issues involved in evaluating the 'specificity' of the effect.

It should be recognized at the outset that learning is not a unitary process. Learning involves a number of components, some of which may be more important during one phase of acquisition than during another. Any investigator attempting to evaluate the specificity of the transfer effect must be cognizant of this fact because the transfer effect may involve transfer of one or more, but not necessarily all, of these components. This consideration makes it extremely difficult to evaluate the specificity of the transfer effect.

Neal Miller (1967) has given a definition of specificity for what he calls 'Grade-A Certified Learning'. Miller suggests that an animal (or group of animals) be trained to emit response R_1 to stimulus S_1 and response R_2 to stimulus S_2. A second animal (or group of animals) is trained to emit response R_1 to S_2 and response R_2 to S_1. He implies that if this test is passed by his experimental animals, everyone will agree that learning has been demonstrated.

Miller's paradigm for demonstrating learning to the exclusion of all other interpretations is excellent and we agree whole-heartedly with it. If one attempts to apply these criteria to the transfer of training phenomenon, however, one immediately discovers a number of problems of interpretation. The Miller paradigm provides an excellent recipe for specificity. One could carry out the training procedure described by Miller, sacrificing animals so trained. Following injection of material from the brain of these animals into naive animals, if one were able to demonstrate that recipient animals receiving extract from the brains of group A donor animals performed as the group A donors did, and that those receiving chemical preparations from the brains of group B donor animals performed as the latter animals performed, one would indeed have demonstrated that the transfer of training effect is specific, and that associations between stimuli and responses had been transferred.

What conclusions can be made, however, if one were able to detect no differences between the two recipient groups? Would the effect simply be non-specific or would it not exist? Let us alter the paradigm slightly by assuming that we train one group of donor animals to emit response R_1 to stimulus S_1 and another group to emit response R_2 to stimulus S_2. We then inject recipient animals with substances extracted from the brains of the donor animals and subsequently test these recipients in the same situation used for donor training. Suppose we discover two things: first, we find that animals injected with material from rats trained to emit R_1 to S_1 do not differ significantly in the responses they emit to S_1 from the animals that have received preparations from donors trained to emit R_2 to S_2. Second, we discover that recipients injected with 'S_1-R_1' tend to emit most of their responses to S_1, whereas those injected with 'S_2-R_2' tend to emit most of their responses to S_2. Unfortunately, in both cases, these are not the 'correct' responses. Still, this appears to be an instance of a transfer of some information about the relevant stimulus, or *stimulus specificity*. Apparently, however, no appropriate information about the response was transmitted via these injections. In any case, we appear to have an instance of another type of specificity, so we should conclude that there are at least two types of specificity – Miller's type (which probably should be called learning) and this second type, or stimulus specificity (which probably should not be called learning).

Let us consider yet a third possibility. Assume we perform for the donor animals the operations specified in the preceding paragraph (training one group to emit R_1 to S_1 and another group to emit R_2 to S_2), and that we subsequently perform the transfer phase of the paradigm. Suppose we then test the injected recipient animals and find that animals injected with 'S_1-R_1' make significantly more R_1's than do animals injected with 'S_2-R_2' and that animals injected with 'S_2-R_2' make significantly more R_2's than do animals injected with 'S_1-R_1'. Let us further assume that there are no differences between our two recipient groups in preference for the particular stimulus pattern to which they respond. That is, they respond indiscriminately to both stimuli, differing only in the responses they emit. As with our second possibility, probably this outcome also should not be considered learning; however, it suggests still another type of specificity, namely, *response specificity*. So now it appears that there are at least three types of specificity: (1) Miller's stimulus–response association (learning), (2) stimulus specificity, and (3) response specificity. Clearly, when one asks whether the transfer effect is specific, one must also specify to which type of specificity one is referring.

An additional problem arises when one realizes that a general type of information may be transferred. Thus, the information involved in the inter-animal transfer effect may consist of a 'focusing' component of learning which manifests itself in the rapid acquisition of the correct response in reinforced learning situations. However, such an explanation, although possibly accounting for data collected in reinforced recipient testing situations, cannot account for results reported in non-reinforced preference tests following administration of material from the brains of trained animals.

What is the most satisfactory way to test for specificity? As the foregoing comments imply, the term itself has not been defined adequately, making it difficult or impossible to undertake the appropriate experiments to test for specificity. There is, however, a variety of approaches to this important question. One of these approaches, suggested by many scientists, has probably received its best formulation by Booth (1967). Booth has suggested that experimental designs involving two situations, differing either in relevant cues or in required activities, should be used. Material from donors trained in situation A would be tested in two recipient groups. One recipient group would be tested in the same situation (Aa), and the other recipient group in the second situation (Ab). Similarly, two additional recipient groups (Ba and Bb) would be used to test for 'transfer' with material obtained from donors trained in the other (B) situation. Booth and others have argued that if the effect is specific, only those recipients tested in situation Aa or Bb should show an effect. There are several problems inherent in this approach, some of which were discussed above.

A second approach to determining the behavioral specificity of the effect has been suggested by Dyal (chapter 13 this volume). Dyal suggests that the appropriate experiment is one in which donors are trained on a discrimination problem in which they learn to respond to a particular stimulus and not to other, similar, stimuli along the same stimulus dimension (conditioned discrimination). If recipient animals injected with trained material are trained to discriminate the stimulus that was 'correct' for their donor, and if they acquire the discrimination significantly faster than control recipients trained with another stimulus along the same stimulus dimension, a relatively strong statement could be made regarding the behavioral specificity of the effect.

This paradigm is probably a more adequate test of specificity than the scheme advocated by Booth and others.

There is yet a third approach, similar to the approach suggested by Dyal. With this third approach, one would attempt to show a correspondence between donor training and recipient transfer effects by demonstrating that variables that affect donor training also affect the transfer phenomenon.

Typically in memory transfer experiments, performance of donor animals is completely ignored, providing that they acquire the response of interest to the experimenter. Terminal performance levels are only rarely reported. With the Dyal approach to specificity, the foregoing is not the case since, by its very nature, the conditioned discrimination paradigm forces the experimenter to pay attention to the course of acquisition of the discrimination in his donor animals.

With the third approach to specificity one would demonstrate that particular variables that are important in the donor training also affect recipient animals in an analogous fashion. If one is able to demonstrate that a variety of these donor learning or performance variables affect recipient performance in a direct way, inferences can be made regarding the specificity of the effect. For example, it is well established that animals receiving overlearning training are less resistant to extinction than are animals trained to an asymptotic level of performance. This is correct even though both groups are not reliably different in terms of their terminal performance during the acquisition phase of such studies. This suggests that one might sacrifice one group following criterion level performance, the other following overtraining, and carry out a transfer study using the brains of these animals. Recipient animals would be trained in the same situation used to train the donors, injected with material extracted from the brains of these donor animals, and subsequently extinguished. Recipient animals receiving material from the overtrained donor group should be less resistant to extinction than recipients injected with material from the criterion-trained donors. Other parameters could be similarly manipulated, and on the basis of a number of independent experiments, a literature would evolve which, when evaluated in toto, would provide strong evidence either for or against specificity.

1.4. Possible chemical mediators

If we assume that a memory transfer effect does in fact exist, we are then left with the problem of determining what chemical or chemicals mediate. Ungar (1969) has maintained rather steadfastly that the chemical involved must be a peptide; indeed, he has recently announced the exact sequence of a 14-amino-acid peptide that appears to transfer 'dark avoidance' from rats (donors) to mice (recipients). In our own experiments, conducted with thousands of animals over the past five years, we have consistently obtained better results with RNA-rich mixtures than we have with the peptide preparation favored by Ungar. So far, however, none of the experiments reported anywhere in the literature offers proof so compelling that an unbiased observer would be willing to consider the matter settled. A brief description of our earlier work will perhaps set the stage for a fuller discussion of the matter.

Early in 1965 we began a replication of the studies reported by Jacobson and his colleagues (Babich et al. 1965). In these experiments, donor rats were trained to approach a cup when a buzzer sounded in order to obtain food. RNA extracted from the donor brains (by roughly the same extraction procedure we describe in this article) was injected i.p. into recipient animals which were then tested in the same apparatus. The buzzer was sounded, but no food was presented to the recipients. Jacobson et al. reported that animals injected with trained material approached significantly more often than did animals injected with untrained material. Although the differences be˯ween groups in our replications of this work were not as great as those found by Jacobson et al., the differences were reliable. We then extended our findings by showing that incubating the trained material with ribonuclease prior to i.p. injection in effect 'wiped out' the transfer effect. A full description of these experiments can be found elsewhere (McConnell et al. 1968).

After successfully replicating this early work by Jacobson and his colleagues, we switched to a different experimental paradigm. Our primary reason for making the change was that we wished to find an experimental situation in which the data could be gathered automatically, by machine, and the use of an operant conditioning chamber and a specific experimental technique called 'magazine training' seemed best to fulfill our requirements. In a series of eight experiments, described in detail elsewhere (McConnell et al. 1968), the donor rats were always trained to approach a food magazine in order to obtain milk. The milk was elevated into the operant chamber by means of a dipper; the dipper remained elevated (and hence accessible to the animal) but a few seconds. If the rat did not get to the dipper in time, it lost the reward. The donor animals soon learned to come to the dipper whenever the solenoid that activated the mechanism sounded. Once learning had reached an asymptote, donors were sacrificed, RNA extraction was performed on their brains and about one-half brain equivalent of the mixture was injected directly into the subdural space above the brains of the recipient animals.

After injection, the recipients also learned to press a lever in the chamber for milk. Ordinarily, rats find it exceptionally difficult to learn such a task, for the dipper is located (in our chamber) on the opposite wall to that from which the lever protrudes. The simple way to train a rat to perform such a task is to undertake the training in two steps. In the first step, the lever is disconnected and the dipper elevated by hand whenever the rat is nearby. The rat then learns to come to the food magazine for milk whenever the solenoid makes its distinctive 'click' sound while elevating the dipper. This step is called magazine training and is, of course, the only training given the donor animals in our experiments. Once the rat has associated the milk with the 'click' sound, it can then progress to the second stage of training, learning specifically to press the lever in order to elevate the dipper itself. We hoped to transfer the magazine training from the donors to the recipients, thus allowing them to skip the first step of the training and learn to press the lever without prior training. Control animals (given injections of untrained material) should be unable to learn the second step, lacking prior magazine training. It was our hope that this paradigm would maximize the differences between animals injected with trained and untrained material since the former should learn the task rather rapidly while the latter simply should not learn at all.

The results of our first experiment are shown in table 1.2. Criterion for learning was 60 bar presses with 30 min of testing. As the table shows, all of the experimental animals (given brain injections of an RNA-rich mixture extracted from trained donors) learned while none of the control animals (given brain injections of an RNA-rich mixture extracted from untrained donors) showed any evidence of learning. A second experiment confirmed this first study, but also suggested that it was critically important that (a) the recipient animals be well-gentled and habituated to handling prior to injection, and (b) that the recipient groups be well-balanced in terms of their pre-injection tendencies to press the lever in the operant chamber.

In the third experiment in this series, we used the same paradigm but added two recipient groups. In this experiment, group A recipients were injected with an RNA-rich mixture extracted from the brains of trained donors,

TABLE 1.2

Number of bar-press responses during each 1 hr daily test session for naive rats.

	Injected with RNA from brains of trained donors					Injected with RNA from brains of untrained donors				
Rat no.	1	2	5	7	10	3	4	6	8	9
Day										
1	11	6	26	9	2	2	2	4	2	0
2	55	13	55	2	46	0	3	0	0	1
3	60	60	60	5	60	3	3	2	1	4
4	60*	34	60	3	60	1	1	0	1	6
5		60*	60*	5	60*	1	4	2	2	4
6				28		1	4	1	0	0
7				60		1	4	2	0	0
8				60		0	4	2	0	0
9				60*		0	1	8	0	2
10						0	3	6	1	4

* Criterion is 60 responses in less than 30 min

group B animals were injected with the same mixture but only after it had been hydrolyzed with ribonuclease, group C rats were injected with an RNA-rich mixture from untrained donors, and group D recipients received an RNA-rich mixture from untrained donors but hydrolyzed with ribonuclease prior to injection. The results of this experiment are shown in table 1.3. As can be seen, hydrolyzation appeared to wipe out the transfer effect.

We have had little success in attempting to use dialyzed materials (such as Ungar does) in the magazine-approach experiments; indeed, we have some tentative evidence that use of dialyzate from the brains of animals trained in an operant conditioning chamber causes what appears to be a 'negative' transfer effect. In one pilot study, we trained donor animals to press the lever to activate the dipper (on a fixed-ratio schedule of one reinforcement for each 50 bar presses); recipients injected with dialyzate from the brains of these trained donors learned significantly more slowly than usual, while animals injected with 'untrained' dialyzate learned normally. Such results are consonant with our findings (in our early attempts to replicate the procedure used by Jacobson and his colleagues) that i.p. injections of materials from trained brains could significantly retard learning when food reinforcement was given the recipients during training, even though a positive transfer

TABLE 1.3

Number of bar-press responses before injection and during each 1 hr daily test session for naive rats.

	Group A Injected with RNA from brains of trained donors				Group B Injected with hydrolyzed RNA from brains of trained donors				Group C Injected with RNA from brains of untrained donors				Group D Injected with hydrolyzed RNA from brains of untrained donors			
Rat no.	14	9	6	2	13	10	17	16	15	11	3	12	1	7	4	5
Pre-injection responses	2	2	4	7	1	2	4	8	1	2	6	6	0	3	4	9
Day																
1	5	8	5	4	0	1	0	4	0	2	3	1	0	1	1	0
2	8	7	3	22	1	2	0	5	0	3	0	3	1	0	1	2
3	60	15	3	60	0	0	0	4	0	1	1	6	1	1	3	16
4	60*	60	60	60*	0	0	0	8	1	2	2	47	0	2	3	19
5		60*	60*		0	1	0	24	0	2	1	60	1	44	4	51

* Criterion is 60 responses in less than 30 min

effect could be demonstrated with the same recipients during unreinforced test trials.

Ungar has argued cogently that peptides may in fact be responsible for the transfer effects we have noted in the experiments above, but his position seems open to dispute. It is true that the RNA mixture we typically use does contain measurable amounts of protein. The actual quantities involved, however, are so small that one would not expect them to be able to mediate an effective transfer, since Ungar (1967) has reported that one typically does not get a transfer effect using less than one brain equivalent of dialyzate (protein). The amount of protein contained in our RNA-rich mixture must be less than one percent of the quantity Ungar finds necessary for an effect. In some of his experiments, Ungar has put homogenized trained brains through a dialysis, then incubated the dialyzate with ribonuclease, trypsin or chymotrypsin prior to injecting it into recipients. Ungar (1968) reported that ribonuclease did not affect the transfer while trypsin or chymotrypsin did. Thus he concludes that RNA cannot be involved in the transfer. However, this conclusion seems untenable, for Ungar adds the ribonuclease to the dialyzate, which can contain no RNA whatsoever. All the RNA would be retained in the dialysis bag, hence Ungar's experiment does not adequately deal with the question of RNA-involvement. Ungar explains our experiments, in which ribonuclease wipes out the effect, by stating that most pancreatic ribonuclease probably contains trypsin as an impurity. However, in some of our experiments we have used dilute NaOH to hydrolyze the RNA rather than RNAse; the wiping out takes place no matter how the RNA is hydrolyzed.

It is our opinion that both RNA and proteins can act as transfer agents. In general, RNA seems to yield better results when injected directly into the brain, while proteins seem more effective when injected i.p. If we consider the question of how memories are stored chemically in the first place, this viewpoint makes an odd sort of sense. The mechanism by which memories are coded or 'stored' in a nerve cell must involve some alteration in cellular metabolism. Since it would be difficult to imagine that a cell could manufacture an altered protein without there being some alteration in the RNA in the cell, one might well assume that memory storage involves a subtle change of some kind in the entire process of protein synthesis. If this is indeed the case, then either RNA or protein (and perhaps even DNA) should be able to act as a transfer agent provided only that it could, upon injection, induce the same subtle change in the metabolism of the recipient as was involved in the brain of the donor animal. At the moment it seems unwise to accept either RNA or protein as being *the* transfer molecule, despite Ungar's recent commendable success in 'decoding' the material presumably responsible for transferring dark avoidance in rats and mice (Ungar 1969). Too many other lines of investigation have suggested that RNA is critically involved in memory storage for this molecule to be dismissed as a potential transfer agent (Zemp et al. 1966, 1967; Hydén 1967; Adair et al. 1968; Shashoua 1968, 1970).

References

ADAIR, L. B., J. E. WILSON and E. GLASSMAN, 1968, Proc. Natl. Acad. Sci. U.S.A. *61*, 917.

AMES, B. N. and D. T. DUBLIN, 1960, J. Biol. Chem. *235*, 769.

ASHWELL, G., 1957. *In:* S. O. Colowick and N. O. Kaplan, eds., Methods in enzymology, vol. 3. Acad. Press, New York, p. 73.

BABICH, F. R., A. L. JACOBSON, S. BUBASH and A. JACOBSON, 1965, Science *149*, 656.

BARONDES, S. H., 1965, Nature *205*, 18.

BOOTH, D. A., 1967, Psychol. Bull. *68*, 149.

BYRNE, W. L., et al., 1966, Science *153*, 658.

CORNING, W. C. and E. R. JOHN, 1961, Science *134*, 1363.

CORSON, J. A. and H. E. ENESCO, 1968, J. Biol. Psychol. *10*, 10.

DYAL, J. A., A. M. GOLUB and R. L. MARRONE, 1967, Nature *214*, 720.

FJERDINGSTAD, E. J., TH. NISSEN and H. H. RØIGAARD-PETERSEN, 1965, Scand. J. Psychol. *6*, 1.

GAY, R. and A. RAPHELSON, 1967, Psychon. Sci. *8*, 369.

GILLESPIE, D. and S. SPIEGELMAN, 1965, J. Mol. Biol. *12*, 829.

GOLUB, A. M., F. R. MASIARZ, T. VILLARS and J. V. MCCONNELL, 1970, Science *168*, 392.

GOLUB, A. M. and J. V. MCCONNELL, 1968, Psychon. Sci. *11*, 1.

GROSS, C. G. and F. M. CAREY, 1965, Science *150*, 1794.

HOFFMAN, R. F., C. N. STEWART and H. N. BHAGAVAN, 1957, Psychon. Sci. *9*, 151.

HUTT, L. D. and L. ELLIOT, 1970, Psychon. Sci. *18*, 57.

HYDÉN, H., 1967, Biochemical changes accompanying learning. *In:* G. Quarton, T. Melnechuk and F. O. Schmitt, eds., The neurosciences. Rockefeller Univ. Press, New York, pp. 765–771.

JACOBSON, A. L., 1967, J. Biol. Psychol. *9*, 52.

KATZ, J. J. and W. C. HALSTEAD, Comp. Psychol. Monogr. *20*, 1.

LOWRY, O. H., N. J. ROSEBROUGH, A. L. FARR and R. J. RANDALL, 1951, J. Biol. Chem. *193*, 265.

LUTTGES, M., T. JOHNSON, C. BUCK, J. HOLLAND and J. MCGAUGH, 1966, Science *151*, 834.

MACHLUS, B. and J. GAITO, 1968, Psychon. Sci. *10*, 253.

MACHLUS, B. and J. GAITO, 1969, Nature *222*, 573.

MCCONNELL, J. V., 1968, Das medizinische Prisma *3*, 3.

MCCONNELL, J. V., R. JACOBSON and B. HUMPHRIES, 1961, Worm Runner's Digest *3*, 4.

MCCONNELL, J. V., T. SHIGEHISA and H. SALIVE, 1968, J. Biol. Psychol. *10*, 32.
MILLER, N. E., 1967, Certain facts of learning relevant to the search for its physical basis. *In:* G. Quarton, T. Melnechuk and F. O. Schmitt, eds., The neurosciences. Rockefeller Univ. Press, New York, pp. 643–652.
REINIŠ, S., 1965, Activ. Nerv. Sup. *7*, 167.
REINIŠ, S., 1969, personal communication.
SCHUTJER, M., 1968, J. Biol. Psychol. *10*, 54.
SEIFTER, S., S. DAYTON, B. NOVIC and E. MUNTWYLER, 1950, Arch. Biochem. *25*, 191.
SHASHOUA, V. E., 1968, Nature *217*, 238.
SHASHOUA, V. E., 1970, Proc. Natl. Acad. Sci. U.S.A. *65*, 160.
TUNKL, J., 1968, J. Biol. Psychol. *10*, 80.
UNGAR, G., 1967, J. Biol. Psychol. *9*, 12.
UNGAR, G., 1969, J. Biol. Psychol. *11*, 6.
UNGAR, G., L. GALVAN and R. H. CLARK, 1968, Nature *217*, 1259.
UNGAR, G. and C. OCEGUERA-NAVARRO, 1965, Nature *207*, 301.
ZEMP, J. W., J. E. WILSON and E. GLASSMAN, 1967, Proc. Natl. Acad. Sci. U.S.A. *58*, 1120.
ZEMP, J. W., J. E. WILSON, K. SCHLESINGER, W. O. BOGGAN and E. GLASSMAN, 1966, Proc. Natl. Acad. Sci. U.S.A. *55*, 1423.

CHAPTER 2

Bioassays for the chemical correlates of acquired information

GEORGES UNGAR

Departments of Pharmacology and Anesthesiology, Baylor College of Medicine
Houston, Texas

2.1. Introduction

There have been three main approaches to molecular mechanisms in the processing of acquired information: direct chemical analysis of the nervous system, use of metabolic inhibitors, and application of biological assay procedures. The first two approaches have suggested that certain chemical changes are correlated with the acquisition and retention of information. There has been a great deal of sterile discussion on the specificity of these correlates for 'learning' as opposed to mere neural activity. The relevance of this debate is doubtful because the definition of learning depends on the interpretation given to this term by various schools of psychology and because learning is inseparable from neural activity. This is the reason why I am trying to avoid the term 'learning' throughout this chapter and substitute for it the broader concept of information processing.

It seems that the real issue is not whether a given chemical change can be correlated with some abstract process of learning but whether there are specific chemical changes corresponding to the processing of each particular set of information. In other words, the question is the existence of a molecular code into which information has to be translated for its processing by the nervous system. Up to the present, only the bioassay methods have attempted to answer this question.

Biological assays represent the only possible approach to a problem in which a chemical process is suspected but the chemical properties of the active material are unknown or inaccessible to direct analytical procedures.

This study was supported by USPH grant no. MH-13361.

Bioassay is the pharmacological method par excellence, but it has found widespread use in many other areas. It is at the origin of all we know about hormones, vitamins and neurohumoral agents. The problem of the molecular coding of information in the nervous system is, in many ways, similar to the problem of chemical mediation of synaptic transmission. When Loewi suspected that the vagal effect on the heart was produced by a chemical mechanism, the only way he could prove it was by 'transferring' the hypothetical chemical mediator from one heart to the other. This is exactly what he did, and thus originated the neurohumoral theory which today dominates our whole thinking about the nervous system.

As is well known, recognition of the chemical mediation of synaptic transmission was preceded by a period of controversy which lasted over a quarter of a century. The similarity to our present problem has been noticed by Schmitt and Samson (1969): 'The notion that transmission at synaptic junctions is mediated not by bioelectric processes but by specific chemical transmitters... seemed speculative, improbable and misleading.... The concept that information-bearing macromolecules may condition the storage and the retrieval of information... seems likely to encounter as much skepticism as did the neurohumoral concept'. Let us hope that, in spite of the far greater complexity of the present problem, the phase of controversy will be shorter. We need additional work in three main areas: we shall have to increase the reliability of the assay method, explore the limits of its specificity and, above all, identify the information-carrying material.

2.2. Experimental data

Although I have been engaged in research on the chemical correlates of neural activity since 1955 (Ungar 1963), the first work of my laboratory dealing more specifically with information processing was started only in 1964. It came as a somewhat unexpected by-product of investigations on the mechanism of drug tolerance. These investigations showed that actinomycin D, an inhibitor of RNA synthesis, could block the development of tolerance to morphine (Cohen et al. 1965). When the work was terminated, about twenty tolerant rats, used as controls, were left over. Instead of destroying them, we decided to use them as donors in a somewhat skeptical attempt to 'transfer' morphine tolerance to recipients which had had no previous experience of the drug. I shall omit these experiments from the present report;

further details are available in Ungar and Cohen (1966) and Ungar and Galvan (1969). In the context of this volume, their significance lies in the encouragement they gave us to attempt other 'transfers' more directly related to the problem of acquired information.

2.2.1. Habituation

The first paradigm chosen to test the possibility of a transfer of acquired behavior was habituation (Ungar and Oceguera-Navarro 1965) because of its similarity to drug tolerance. In both cases, the animal learns to suppress an innate response to a chemical or physical stimulus. In our first experiments, donor rats were habituated to a loud sound which originally elicited a startle response. After repeated presentation of the stimulus, the startle responses were reduced to 10% or less. Under our experimental conditions, this took about 10 to 14 days. Extracts of brain taken from these donors were injected into naive recipient mice. When these were tested, 24 hrs later, they showed a significant reduction in their startle responses as compared with the controls, which were injected with extracts of brain taken from untrained donors. These experiments were first presented in April 1965 at a symposium on Proteins of the Nervous System in Bruges (Ungar 1966).

The results of these experiments were criticized mostly on the ground that habituation is not considered to be real learning by some schools of psychologists. Whether we call it learning or not, habituation is certainly the result of the acquisition and retention of information. I was more concerned with the fact that this type of experiment involved training of the recipients, since every time they were tested they were further habituated. This objection can also be raised for some other paradigms used subsequently (Ungar 1967a, b) and which, for this reason, are omitted from this report. I shall mention only those experiments in which the recipients were never reinforced so that their behavior could be influenced only by whatever information was contained in the brain extract they received.

2.2.2. Escape training

These experiments were done in a Y-maze provided with a grid floor and lights and buzzers at the end of each branch of the Y. The donor rats were placed in the stem of the Y and given electric shocks; they were trained to escape either on a brightness discrimination clue (lighted or unlighted arm of the maze) or on spatial discrimination (left or right arm). A more complex paradigm involving auditory and visual clues will be described below (section 2.4.2).

We did a large number of experiments with brightness discrimination (Ungar and Irwin 1967; Ungar 1967b), and escape into the lighted arm of the maze showed a high degree of reproducibility (table 2.1). The results depended, however, on two critical factors: the length of training given to the donors and the dose of brain administered. It should be noted that training the donors to escape into the unlighted arm of the maze could not be transferred.

TABLE 2.1

Action of brain extracts from donors trained to escape to lighted arm (L+) and to unlighted arm (L−), and from untrained rats on responses of naive, unreinforced recipients.

	% Escape to light		
	Experimental		Control
	L+	L−	
Preinjection mean	57.2	56.8	57.5
Postinjection* mean	69.8	62.3	60.2
N**	56	12	40
P†	<0.01	>0.05	>0.05

* Received amounts equivalent to 0.75 to 1.5 g of donor brain

** Number of recipients

† Computed by U test

Transfer of escape into the left or right arm gave much less consistent results. Some of the experiments were remarkably successful (table 2.2) but, on the whole, reproducibility was poor. Not only were the extracts sometimes ineffective but occasionally the results were reversed; i.e. recipients of right-trained brain turned predominantly to the left and vice versa. We tried to analyze the various factors which determined the behavior of the recipients and found that the left or right bias of the donors and of the recipients may have had a decisive influence on the outcome of the experiments. I concluded that the main message we transmitted to the recipients was not to turn left or right but to overcome their bias (Ungar 1967b). Because of these complicating factors, we abandoned this type of experiment.

TABLE 2.2

Action of brain extracts from donors trained to escape to the left (L), right (R) and untrained rats (C) on escape behavior of unreinforced recipients.

	% Left turns in maze		
	L	R	C
Preinjection mean	38	51	48
Postinjection day 1	65*	28	50
day 2	58†	28	49
day 3	41	25**	47
day 4	45	28	48
day 5	49	29	49
Postinjection mean	51	28**	48
N††	6	6	20

*$p < 0.001$
**$p < 0.01$
†$p < 0.05$. All comparisons were made with preinjection mean by U test
†† Number of animals

2.2.3. Passive avoidance

In 1967, Gay and Raphelson reported the transfer of acquired dark avoidance. They trained rats to overcome their natural preference for the dark by giving them electric shocks when they ran into the dark chamber of a three-compartment system linked by passages. We were able to confirm their results and adopted the paradigm with some minor modifications (Ungar et al. 1968). It offered several advantages over the other procedures tried previously: it was more easily reproducible, the effect was clearly dose-related and both training and testing were less time-consuming. We determined the optimum conditions for the training and used a number of control situations to prove that the dark avoidance observed in the recipients was not due to the stress produced by the repeated administration of electric shocks (table 2.3).

Another passive avoidance paradigm, the avoidance of step-down from a platform, was also found to be transferable. When rats or mice are placed on a narrow platform, within a few seconds they tend to step down to the wider area below it. When, however, they are subjected to electric shocks in this area, the latency of stepping down increases considerably. Table 2.4 summarizes these experiments. We have obtained consistent results with step-down avoidance but the effect was not as clearly dose-related as in dark avoidance. We also experienced difficulties in automating the testing procedure. We decided, therefore, to use dark avoidance as the standard paradigm for chemical identification of the active material.

TABLE 2.3

Effect of brain extracts from dark-avoidance trained donors and variously treated control donors on time spent in the dark box (DBT) by the recipients.

Training of donors	DBT*		N**
	%	± S.D.	
Dark avoidance	22.5	10.5	32
None	66.7	11.3	24
Run to dark box but no shock	78.0	17.8	12
Shocked in lighted chamber	68.3	11.1	12
Shocked in white box	51.6	18.8	8

All recipients were injected with the equivalent of 1 g donor brain. They were tested 1, 2 and 3 days after injection and the mean of the three tests obtained in each animal was used to compute the values given in the table

* 100% = 180 sec
** Number of animals

TABLE 2.4

Avoidance of step-down from platform.

	Latency sec ± S.D.		Total time on platform sec ± S.D.		N
First series					
Control	18	12	–	–	20
Experimental	45	16	–	–	32
Second series					
Control	5	2.5	16	7	60
Experimental	19	8	34	15	86

All recipients were injected with the equivalent of 1 g donor brain. They were tested 1, 2 and 3 days after injection and the mean of these tests obtained in each animal was used to compute the values given in the table

2.3. *Reliability of the assays*

One of the obstacles to more widespread acceptance of the bioassay approach has been the difficulty experienced in some laboratories to replicate the results. This was particularly true at the early stages when the conditions for successful transfer were almost completely unknown. Even under the best of conditions, bioassays are notorious for their low level of reliability and accuracy and the evaluation of their results requires careful statistical analysis.

2.3.1. Conditions for successful transfer

Since the original publications in 1965, the literature of the bioassay approach has been rapidly growing. As of November 1969 there were at least 81 publications on the subject, 69 of them reporting successful transfer experiments from 22 laboratories. I undertook, therefore, a survey of the literature with the purpose of analyzing the factors which determine success or failure. Details of this survey have been given in another publication (Ungar 1970a) and the following is only a brief summary of my evaluation of the conditions of success.

Adequate donor training. This point is still often misunderstood because it is not realized that the bioassay requires a high concentration of the information-carrying material in the brain of the donors. It is not sufficient for the animal to learn the task to be transferred; time must be allowed for the chemical correlate to be synthesized and accumulated. We analyzed this factor in the case of dark avoidance (Ungar et al. 1968). It is important that the donors be as fully trained as is compatible with the difficulty of the task. Inadequately trained or untrained animals were poor donors in our as well as Reiniš' experience (1966). On the other hand, overtraining beyond a certain point decreased the quality of the donors (Ungar and Irwin 1967; Ádám and Faiszt 1967; Ungar et al. 1968). It seems that during training there is a burst of synthesis of the information-carrying molecules which, as soon as the memory is consolidated, tends to subside. A well-learned behavior may require only a minimum amount of this material in the brain for retrieval.

Preparation of the extracts. In the absence of any knowledge of the nature of the material involved in information processing, the logical procedure is to start with a crude extract and purify it stepwise under the guidance of bioassays. This was not done in the majority of the studies where preparation was aimed at isolating an RNA fraction of brain. I shall return to this point in section 2.5 because it may have been quite critical in the attempts at replicating the experiments of Babich et al. (1965). Whatever the nature of the information-carrying substance is assumed to be, there are elementary rules for the preparation of biologically active compounds. If these rules are not observed, the material may be lost during the extraction procedure. It can be safely stated that for any organic molecule present in living systems there is an enzyme capable of destroying it. In living tissues the molecules are often protected but as soon as the tissue is ground up, the barriers that normally separate enzymes and substrates break down. It is, therefore, essential to operate at low temperature throughout the extraction and to keep the preparation frozen. Even in the absence of active enzymes, the stability of the material may depend of pH, ionic strength of the solution, possibility of oxidation and other factors.

Dose of brain administered. In our experiments this point was extremely critical. Successful assays always required administration of at least twice the equivalent of the recipients' brain. Since we usually used rats as donors and mice as recipients, it would be unwise to generalize from our experience. However, with one exception (Rosenblatt 1970), all the positive transfer experiments used the same high doses. On the other hand, in the negative experiments analyzed in my survey (Ungar 1970a), 83% of the animals used received the equivalent of only one brain or less.

Selection of recipients. As a general rule, one should make sure that the recipients (the species, strain, age or sex used) are able to learn by training the task to be transferred to them. Too young or too old animals make poor recipients. Even if the recipients are not reinforced, as in most of our experiments, they should be properly motivated. Reiniš (1966) observed that food-rewarded behavior can be transferred only to hungry recipients. One must also make sure that the recipient is suitable for demonstrating a given response. For example, in the assay for dark avoidance, it is essential that, before receiving the extract, the recipients show a marked preference for the dark.

Schedule of testing. This extremely important factor has often been neglected. The interval after injection of the extract at which the effect can be observed varies with the situation. It is, however, seldom less than 24 hrs and often reaches two or three days. Testing at a single arbitrarily chosen interval, as was often done in the negative experiments, cannot justify a negative conclusion. Testing must be done at several intervals spaced between 6 hrs and 5 or 6 days.

Blind testing. It is essential that the personnel testing the recipients be kept ignorant of the treatment received by the animals. Experimental and control animals should be mixed and handled under code numbers.

Trial and error for optimal conditions. The factors just summarized are those that I found important in the experiments done in my laboratory. Under different conditions, they may act differently. In each particular situation, the best conditions have to be found by trial and error.

2.3.2. Statistical evaluation

Even the simplest bioassays, using isolated tissues or organs, have a limited reliability. It is further reduced when the test object is a whole organism. This is true even when the test is based on simple responses such as a change in blood pressure, in blood chemistry or in some morphological feature. It can be assumed that an assay utilizing a behavioral response would have the lowest predictability of all.

In spite of this serious handicap, the method can be made to work if the results are submitted to statistical evaluation. A minimum of familiarity with biological assay techniques and common sense suggest that not all individual experiments will be statistically significant. The validity of the results can be established by computing the overall reliability of pooled data. Table 2.5 summarizes the results of several series of experiments obtained in my laboratory between November 1964 and October 1967. Each of the series yielded reliable differences between the control and experimental groups. This was true even for the left–right discrimination paradigm in which many experiments were negative or reversed. The overall probability that the effect of trained brain extracts was due to chance is less than 1 in 10,000.

Since 1967, we have been concentrating on the passive avoidance paradigm, especially dark avoidance. To date (November 1969), 115 extracts have been prepared, each from an average of 50 trained rat brains. During the same period, 34 control extracts were made. Table 2.6 shows the distribution of the activity of these extracts in terms of the time spent in the dark box (DBT) by the recipients injected with them. It indicates the probability by the χ^2 test when the results were divided into two groups; those extracts that reduced DBT to a level below 50% and those that failed to achieve this. The

TABLE 2.5

Behavioral changes in animals injected with extract of brain taken from trained (TRB) or untrained (NRB) donor rats (1964–1967).

	Number of recipients				p[2]
	TRB		NRB[1]		
	+	0 or −	+	0 or −	
Habituation	41	1	0	22	<0.001
Conditioned avoidance					
Reinforced	10	5	2	11	<0.01
Escape to light					
Reinforced	15	9	5	15	<0.02
Unreinforced	80	12	2	38	<0.001
Left or right escape	110	67	2	28	<0.001
Audio-visual					
discrimination	30	9	2	14	<0.001
Totals	286	103	13	128	<0.001
%	73.5	26.5	9.2	90.8	
N	389		141		

\+ = Change in the direction expected from the training of donors
0 = No change; — = change in opposite direction

[1] In control animals, the direction of the change was counted as in the corresponding experimental animals
[2] χ^2 method

overall mean time spent in the dark box by recipients treated with trained extracts was 34.6%±9.6 and in those treated by control extracts it was 66.7%±11.1. The preinjection DBT of the whole population was 72.6%±8.3. The *t* test indicated less than 1:10,000 probability of the controls belonging to the same population as the experimental recipients.

TABLE 2.6

Distribution of dark-avoidance trained and untrained brain extracts prepared during the period between October 1, 1967 and November 1, 1969, according to their effect on time spent in the dark box (DBT) of the recipients.

DBT (sec)	Experimental		Control	
	No	%	No	%
20– 29.5	3	2.6	0	0
30– 39.5	7	6.1	0	0
40– 49.5	16	13.9	0	0
50– 59.5	26	22.6	0	0
60– 69.5	23	20	1	2.9
70– 79.5	23	20	2	5.9
80– 89.5	10	8.7	0	0
90– 99.5	5	4.3	2	5.9
100–109.5	2	1.7	5	14.7
110–119.5	0	0	5	14.7
120–129.5	0	0	5	14.7
130–139.5	0	0	7	20.5
140–149.5	0	0	7	20.5
Total	115		34	
<90	108	93.9	3	8.8
$\geqslant 90$	7	6.1	31	91.2

$\chi^2 = 101$; $p < 0.0001$

2.4. *Specificity of the assays*

One of the most important problems that remain to be solved is the specificity of the assay method. Experiments using unreinforced recipients clearly indicate that the brain extracts do not act merely by stimulation of learning. We have done experiments in our laboratory on the stimulus specificity of habituation, escape behavior and passive avoidance. Their results suggest that the effect has a large degree of specificity as far as the sensory modality of the stimulus and the gross characteristics of the response are concerned. What is still missing is evidence for finer specificities within the same sensory modality (color, sound frequency, smell, taste etc.) or for well-defined motor responses. A good beginning is represented by the experiments of Zippel and Domagk (1969) described in chapter 10 of this volume.

2.4.1. *Habituation*

We showed the stimulus specificity of the habituation assay by submitting one group of donors to the sound stimulus, as mentioned above (section 2.2.1), and another group to an air puff. When the 10% criterion was reached, extracts of brain from sound-habituated donors, air puff-habituated donors and untrained rats were injected into groups of recipients which were tested for sound habituation. As seen in fig. 2.1a, only the recipients of sound-habituated brain showed loss of startle responses to sound. Conversely (fig. 2.1b), when three similarly treated groups were tested with air puff, only those that received air puff-habituated rat brain showed habituation to air puff (Ungar 1967a).

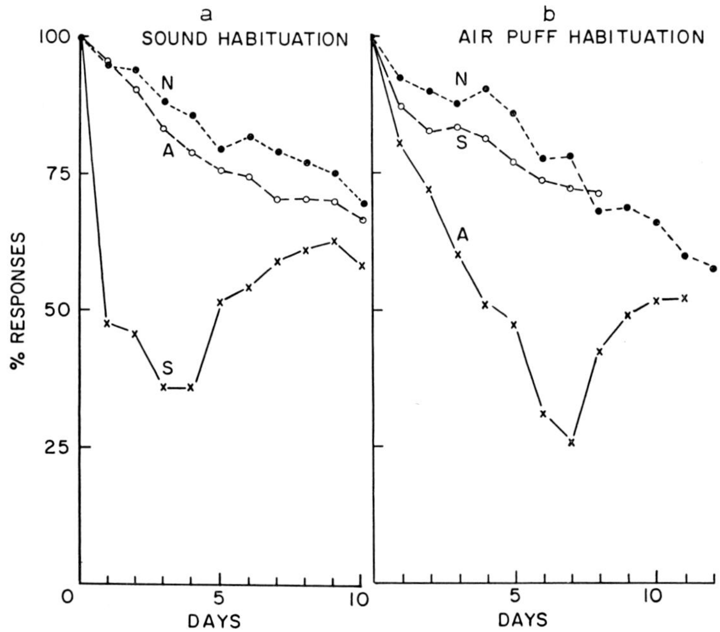

Fig. 2.1. Stimulus specificity of habituation assay. Abscissa: days of testing after injection; ordinate: number of startle responses to 100 stimuli. (a) Responses to sound in mice injected with normal (N), air puff-habituated (A) and sound-habituated (S) rat brain. (b) Responses to air puff in similarly treated groups of recipients (Ungar 1967a).

2.4.2. Audio-visual discrimination

Stimulus specificity of escape training was tested by training one group of donors to escape to the left on a light signal and to the right on the sound of a buzzer. Another group was trained in the opposite direction by the same two signals. Fig. 2.2 shows that the recipients did respond to the signals predominantly in the same way as the trained donors whose brain was administered to them.

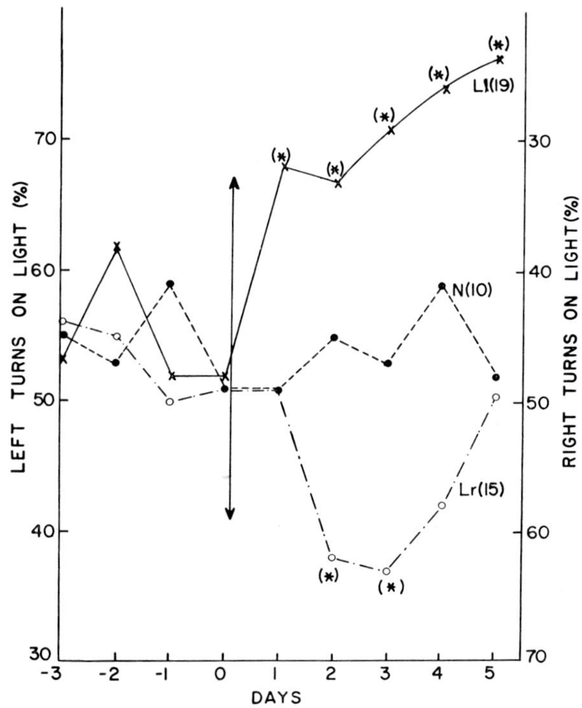

Fig. 2.2. Y-maze performance of mice injected with brain extracts taken from donors trained to turn left on light and right on buzzer (Ll), donors trained to turn right on light and left on buzzer (Lr) and untrained rats. Abscissa: days before and after injection of 1 g of brain; ordinate: left, left turns on light (or right turns on buzzer), right, right turns on light (or left turns on buzzer). Numbers between parentheses indicate number of recipients in each group.

2.4.3. Cross-transfer of passive avoidance

In the first experiment, one group of recipients injected with extracts of brain from donors trained for dark avoidance, another trained for avoidance of step-down, and a control group given untrained brain were tested for dark avoidance. Fig. 2.3 shows that only those recipients which were treated with brain

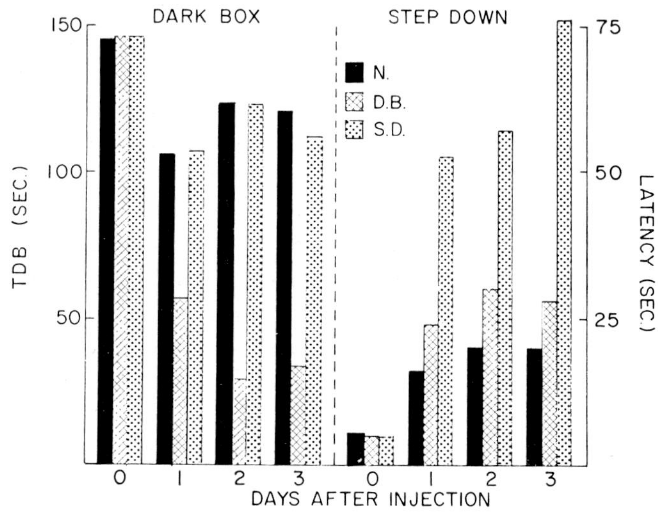

Fig. 2.3. Cross-transfer between dark avoidance and step-down avoidance. On left, time spent in dark box (DBT) by recipients injected with brain extracts from untrained (N), dark-avoidance trained (D.B.) and step-down-avoidance trained (S.D.) donors. On right, latency of step-down in similarly treated recipient groups (Ungar 1969).

from dark-avoiding donors had a reduced DBT. Conversely, when similarly treated groups of recipients were tested for step-down avoidance, the latency of step-down increased only in the group given brain from step-down avoiding donors (Ungar 1970d).

All three experiments clearly indicate that the information transferred includes specific clues as to the modality of the stimulus, sound, air-pressure, light or dark, spatial position etc. Experiments are now being planned for finer specificities.

2.5. *Chemical identification of the information-carrying molecules*

Our main purpose in this laboratory is the isolation, purification and identification of the factors extracted from donor brain and the bioassay is only a tool which serves to guide us in the chemical fractionation.

In all the experiments mentioned in the previous sections, the first assays were done with homogenates. When these were active, the homogenate was centrifuged (at 25,000 to 65,000 *g*) and both the supernatant and the resuspended sediment were tested. In all cases, the activity was in the supernatant. The next step was dialysis aimed at separating the large from the small molecules at the approximate level of 10,000 M.W. Activity was found in the dialyzate for morphine tolerance, habituation, left–right discrimination, dark avoidance and step-down avoidance. The factors transferring light–dark discrimination and conditioned avoidance in the Y-maze (Ungar 1967a) were found to be non-dialyzable and no further attempts were made for their isolation.

The dialyzable factors were concentrated by lyophilization and further purified by addition of 20 vol of cold acetone. All of them were precipitable by acetone. The redissolved precipitate was fractionated by gel filtration on Sephadex-G50 and -G25, ion exchange resins and thin-layer chromatography. The most highly purified material is the factor isolated from the brain of dark-avoiding rats; it is active at 0.1 μg per mouse. It occupies a single chromatographic spot, stained with ninhydrin. This spot is absent in brain extracts of naive rats. This factor has been identified as a pentadecapeptide named scotophobin. The amino acid sequence, established by mass spectrometry, is as follows: Ser-Asp-Asn-Asn-Gln-Gln-Gly-Lys-Ser-Ala-Gln-Gln-Gly-Gly-TyrNH_2. The sequence has now been confirmed by synthesis by Dr. W. Parr at the University of Houston.

An important step in the identification of the materials was a study of their susceptibility to enzymes. None of the factors investigated was destroyed by pancreatic ribonuclease and all of them were inactivated by a protease, either trypsin or chymotrypsin or both (table 2.7). These results, together with the chemical properties, suggest that the active materials are peptides or at least that their activity is dependent on some intact peptide linkages.

The majority of workers who attempted to transfer acquired information assumed that the active substance was an RNA sequence. This assumption

TABLE 2.7

Action of proteases and pancreatic ribonuclease on the activity of four transfer factors.

Factor	Destroyed by		
	Trypsin	Chymotrypsin	Ribonuclease
Morphine tolerance	0	+	0
Sound habituation	0	+	0
Dark avoidance	+	0	0
Step-down avoidance	+	+	0

was at the origin of a great deal of misunderstanding and was probably an important factor in the failure of the early attempts at reproducing some of the experiments. Babich et al. (1965) used an RNA preparation which was certainly heavily contaminated with proteins and other materials. Some of the several groups of workers who tried to replicate their results were more competent in the preparation of brain RNA and, thus, could have eliminated the active principle.

Dr. Fjerdingstad, who worked in my laboratory in 1968–1969, and I tried to resolve the contradiction. We used the same pool of brains taken from donors trained for dark avoidance. He prepared an RNA extract according to the procedure he and his coworkers used in Denmark (Røigaard-Petersen et al. 1968). I prepared an extract following the procedure outlined above. Table 2.8 shows that both preparations were active and their activity was destroyed by trypsin but not by ribonuclease. These results suggested that the active substance may be a peptide forming a complex with RNA. The probability of such a complex being formed was supported by the basic characteristics of the peptide. The hypothesis was confirmed when the RNA preparation was dialyzed at different pH levels. From pH 8 to 6.5 the activity remained in the retentate but between pH 6 and 5.25 it appeared in the dialyzate. Between pH 5 and 4.5 the factor became again non-dialyzable. This was explained later by the fact that it forms another complex with some acidic material which is present in the dialyzate and which has an isoelectric point in the vicinity of pH 4.75. From pH 4 down, the dark avoidance factor is freely dialyzable. Preparation of the factor by passing through the RNA-bound step presented some distinct advantages and we have adapted it for our routine isolation technique.

Up to now, comparison between the two methods of preparation has been

TABLE 2.8

Comparison between 'peptide' and 'RNA' preparations of brain from dark-avoidance trained donors on time spent in the dark box (DBT) by recipients.

	DBT (sec)*	± S.D.
'Peptide'	76	26
'RNA'	67	21
'RNA' and trypsin	109	29
'RNA' and chymotrypsin	68	34
'RNA' and ribonuclease	72	30
Control	121	25

* Means of 12 animals tested 24 to 48 hrs after i.p. injection of extracts

done only for the dark-avoidance factor. Until it is done for other active substances, we may be allowed to suppose that they also are peptides that have a tendency to form complexes with RNA and can be extracted with it.

2.6. *Tentative interpretation*

It is no longer possible to ignore the results of biological assay procedures in any speculation on the molecular mechanism of neural information processing. Even without knowing the extent of the specificity of the extracts, it seems reasonably certain that the chemical correlates of learning do not represent merely a non-specific stimulation of protein synthesis necessary for the increased production of synaptic transmitters or for neuronal growth. The results of bioassay clearly indicate that the molecules whose increased synthesis has been shown by direct analysis of the brain and whose essential role has been demonstrated by the use of metabolic inhibitors do actually take part in the coding of acquired information.

Only two types of interpretations remain compatible with the facts: the so-called 'tape recorder' molecule hypothesis and the theories based on the chemospecificity of neural pathways. For further examination of these hypotheses see the reviews by Ungar (1970b, c). I should just like to outline here the interpretation which I have proposed (Ungar 1968) and which represents the working hypothesis of this laboratory.

The work of Sperry (1963), Jacobson (1969) and others has suggested that the existence of a molecular recognition system is the most probable explanation of the mechanism by which the nervous system is organized during embryonic life. Molecular labeling would allow the neurons which are to form synaptic connections to recognize each other. The possible details of the process have been worked out by Roberts and Flexner (1966) by means of ingenious assumptions.

In the light of this general idea, one can assume that at birth the brain is provided with a built-in molecular code, the same code which controlled the differentiation and organization of the pathways and centers according to a plan laid out in the genome. One can suppose that this code can be adapted to the postnatal needs for processing acquired information and for adapting behavior to the constantly changing conditions of the environment. In other words, the innate system could reprogram itself using the genetically organized pathways.

The mechanism by which this reprogramming takes place is essentially the formation of new synaptic junctions between existing pathways which may be brought about by the process of simultaneous firing or concurrent activity. It is probable that when two neurons in close proximity fire simultaneously or successively in a definite sequence, the permeability changes taking place at the point of contact can establish a connection between them. In view of the extremely rich contacts between neurons in the brain, the probability of such connections being created is quite high and is a function of the complexity of the innate organization.

The new synapses created in any given learning paradigm are chemically encoded in a molecular complex formed by the labels of the pathways involved. A new synapse formed between a pathway marked by code-name *a* and another marked by code-name *b* will be adequately characterized by the complex *ab*. The formation of such a complex takes place probably at thousands of similar synapses so that when we extract the brain of a donor, we extract significant amounts of substance *ab* which, when injected into the recipients, reconstitutes in their brain the synapse coded by this name. For detailed discussion of the hypothesis, see the original article (Ungar 1968) and the reviews mentioned above (Ungar 1970b, c).

It is impossible to tell at present whether this hypothesis is anywhere near the truth. I believe, however, that it will become accessible to experimental verification when some of the information-carrying molecules are identified and can be isotopically labeled. We shall then be able to follow their destiny after injection into the recipients and to pin-point the structures with which they become associated.

In any case, the biological assay method has already made notable contributions to our understanding of the workings of the brain. It is to be expected that in a very near future it will become the most important approach to an area of research that is considered by many as the last frontier of biology.

References

ADAM, G. and J. FAISZT, 1967, Nature *216*, 198.

BABICH, F. R., A. L. JACOBSON, S. BUBASH and A. JACOBSON, 1965, Science *149*, 656.

COHEN, M., A. S. KEATS, W. KRIVOY and G. UNGAR, 1965, Proc. Soc. Exp. Biol. Med. *199*, 381.

GAY, R. and A. RAPHELSON, 1967, Psychon. Sci. *8*, 369.

JACOBSON, M., 1969, Science *163*, 543.

REINIŠ, S., 1966, Worm Runner's Digest *8*, 7.

ROBERTS, R. B. and L. B. FLEXNER, 1966, Am. Scientist *54*, 174.

RØIGAARD-PETERSEN, H. H., T. NISSEN and E. J. FJERDINGSTAD, 1968, Scand. J. Psychol. *9*, 1.

ROSENBLATT, F., 1970, Induction of behavior by mammalian brain extracts. *In:* G. Ungar, ed., Molecular mechanisms in learning and memory. Plenum Press, New York, 103–147.

SCHMITT, F. O. and F. E. SAMSON, Jr., 1969, Neurosci. Res. Prog. Bull. *7*, 277.

SPERRY, R. W., 1963, Proc. Natl. Acad. Sci. U.S. *50*, 703.

UNGAR, G., 1963, Excitation. Thomas, Springfield, Illinois.

UNGAR, G., 1966, Configurational and hydrolytic changes of proteins on excitation; their role in information processing. *In:* H. Peeters, ed., Protides of the biological fluids, Bruges, 1965. Elsevier Publishing Company, Amsterdam, pp. 151–162.

UNGAR, G., 1967a, Chemical transfer of acquired information. *In:* Proceedings of the Fifth International Congress CINP, Washington, D.C., 1966. Excerpta Medica, Amsterdam, pp. 169–175.

UNGAR, G., 1967b, J. Biol. Psychol. *9*, 12.

UNGAR, G., 1968, Perspectives Biol. Med. *11*, 217.

UNGAR, G., 1970a, Chemical transfer of acquired information. *In:* A. Schwartz, ed., Methods in pharmacology. Appleton-Century-Crofts, New York. Vol. 1, pp. 479–513.

UNGAR, G., 1970b, Role of proteins and peptides in learning and memory. *In:* G. Ungar, ed., Molecular mechanisms in learning and memory. Plenum Press, New York, pp. 149–176.

UNGAR, G., 1970c, Intern. Rev. Neurobiol. *13*, 223.

UNGAR, G., 1970d, Chemical transfer of learned behavior. *In:* A. Lajtha, ed., Symposium on protein metabolism in the nervous system. Plenum Press, New York, pp. 571–585.

UNGAR, G. and M. COHEN, 1966, Intern J. Neuropharm. *5*, 183.

UNGAR, G. and L. GALVAN, 1969, Proc. Soc. Exp. Biol. Med. *130*, 287.

UNGAR, G., L. GALVAN and R. H. CLARK, 1968, Nature *217*, 1259.

UNGAR, G. and L. N. IRWIN, 1967, Nature *214*, 453.

UNGAR, G. and C. OCEGUERA-NAVARRO, 1965, Nature *207*, 301.

ZIPPEL, H. P. and G. F. DOMAGK, 1969, Experientia *25*, 938.

CHAPTER 3

The validity and reproducibility of the chemical induction of a position habit

W. F. HERBLIN

Central Research Department, E. I. du Pont de Nemours and Co., Wilmington, Delaware

3.1. Introduction

The proposition that memory and/or learning can be transferred from one animal to another is certain to be controversial. The first objection will come from those who hold that the higher processes of the mind are beyond the realm of the physical sciences and should be examined only by psychic methods. The next objection will come from those who disbelieve anything they cannot understand and demand a unifying theory. A final group of objections will be directed toward real or imagined defects in the logical or statistical design of the experimental program. Therefore, it should not come as any surprise that the first planarian transfer study started an argument which has still not been resolved.

Recently, several symposia and review articles have adequately summarized the existing evidence for 'memory transfer' as well as the arguments against it (Byrne 1970; Rosenblatt 1970a). I defer to these authors, who have performed a monumental task, and I will not attempt any sort of review. I will also avoid the lengthy discussion of arguments for and against by simply ignoring them. I will not even spend any time discussing the relative merits of the various names by which the phenomenon has been called. All of these things have been discussed at length elsewhere.

My only purpose in preparing this chapter is to discuss a few logical and statistical problems which are involved in the investigation of the memory transfer phenomenon. Since I will use some of my own data to support or refute certain points, I will describe our methods in some detail. This should not be interpreted as an exclusive recommendation of our method, but as an attempt to use data with which I am most familiar. As will be seen later, the data are reported as probability values. I feel that this deserves some comment and the topic will be included in section 3.2. Once these processes of data accumulation and tabulation are defined, we can proceed to examine certain objections and contentions which bear on the reproducibility and/or validity of the effect we are measuring.

3.2. Methods

Our methods were originally as identical to those of Rosenblatt et al. (1970b) as we could make them. As the experiments progressed, however, and as we proceeded beyond a simple replication, various modifications were found necessary. I will describe certain aspects of the method in some detail here as they are essential to the subsequent discussion. The full procedure has appeared elsewhere in the literature (Rosenblatt 1970b).

3.2.1. Donor training and extract preparation

Female Holtzman rats (60–75 days old) were food deprived for 24 hrs and then placed on a restricted feeding schedule to maintain their weight at 80–90% of their initial weight. The training box was a standard rodent training cage equipped with two levers on one end and a food cup on the other. A treadle occupied the half of the box containing the food cup. The other half was the standard grid floor. The donor animals were required to learn to press either the right or left lever for food reward. A response on either the correct or the incorrect lever inactivated both levers and a micro-switch under the treadle had to be tripped to reactivate the reward circuits. Therefore, the animals also had to learn to cycle between the ends of the box.

It was found that 150 min of training administered as either ten 15-min sessions or fifteen 10-min sessions resulted in well trained animals. The correct lever was rewarded with 45 mg Noyes pellets but since only one response was allowed per cycle, some presses on the correct lever were not rewarded. After the 150 min of training, the donor animals were sacrificed by decapitation and the extract prepared.

Unless otherwise stated, the extract used in our studies was that fraction of an aqueous extract of whole brain which precipitated between 50 and 75% acetone at 0–4 °C.

3.2.2. Recipient handling

Recipient animals underwent three separate procedures arbitrarily called preconditioning, pretest and test. Since I will frequently refer to these procedures and the results obtained from them, it is important that the similarities and the differences are clearly understood.

Preconditioning refers to training in a single bar box. One lever is removed and the other lever is centered in the end of the box. This is, in effect, the removal of the incorrect lever. The animal must learn to press the lever for food reward and must also learn to cycle from one end of the box to the other. The amount of preconditioning is critical and is controlled by criteria of both time and presses. Each animal is given five sessions consisting of 10 min or 25 cycles, whichever occurs first. The evolution of this procedure is discussed later. The purpose of the preconditioning is to provide a rat which will show high, unbiased performance when placed in the two-lever box.

Pretest refers to any sessions given to a rat under test conditions before injection. This means that the rat is placed in the two-lever box under conditions of universal reinforcement. Normally this amounts to two 5-min sessions, 4 hrs apart, on the day of injection. This procedure provides a baseline for each rat and measures initial bias. That initial bias exists is shown by the fact that a small percentage of the rats will score over 90% on one lever in the first pretest session. Since they are rewarded for these responses, they usually maintain this bias throughout the subsequent test sessions.

Test sessions are those sessions in the double-lever box which follow injection. Universal reinforcement conditions exist as does the cycling requirement. Test sessions are usually two 5-min sessions per day, 4 hrs apart, continued for three days after injection.

3.2.3. Data reduction

Ignoring the preconditioning data, each rat in each session produces two numbers which are immediately converted to a percent left score. This conversion is made according to the formula

$$\frac{\text{Left presses} + 1}{\text{Total presses} + 2} \times 100$$

The formula has little effect if the number of total presses is high and tends to drive lower scores toward 50%. Any score which is based on less than two presses is eliminated. In fact, however, the preconditioning procedure effectively eliminates low scores and since its adoption, less than 1% of the scores have fallen into this category. These percentage scores are then used to calculate an average pretest score and an average test score for each rat. These eight scores (average pretest, tests 1–6, and average test or mean) make up the raw scores for each rat and are used for statistical analysis.

The next step is the computation of derived scores. After several possible methods were tried, we adopted the rather simple procedure of measuring the deviation of each test score from the average pretest. The six deviation scores are then averaged to provide a mean deviation. These derived scores are also subjected to statistical analysis.

3.2.4. Statistical analysis

Although we have routinely applied a variety of statistical procedures to our data in this age of computers, I will only discuss and report results from one technique: the Mann–Whitney U test (Siegel 1956). I should add that in most cases the various statistical procedures agreed quite well.

We used the U test to obtain a measure of the difference between two groups. In all cases we had two sets of scores, whether raw or derived. One set was from animals assigned to the right group and the other set was from animals assigned to the left group. By calculating U and assuming its normal distribution, we could obtain a value of p, the probability that our two sets of scores came from the same population. We used single-tailed probabilities and I feel this was justified. However, it would make no difference whatsoever in the general conclusions to use two-tailed probabilities. We have tabulated these probabilities such that a value of $p < 0.50$ indicates that the groups differed in the predicted direction and a value of $p > 0.50$ indicates an inversion.

Control animals were treated in a completely analogous manner. The only difference between control groups and experimental groups was in the injection procedure. Some control animals were not injected, some were injected with saline, and some were injected with extracts from untrained donors. We observed no differences among these various control groups and have pooled the data. It is important to note, however, that each control animal had been randomly assigned to either the left or the right and these designations provided left and right control groups which were identical with the left and right experimental groups except for treatment at injection.

3.2.5. Data tabulation

Once the experiment has been designed and performed, some method must be found to adequately convey the results to others. The selection will depend in part on how much it is desirable to convey and somewhat on the general acceptance of the results and particularly the conclusions. A logical extension of knowledge will be accepted on less evidence than a result which contradicts previous reports. In the case of memory transfer, which many feel is illogical at the start, the data must be reported more completely than usual. Simple statements of statistical significance are not sufficient.

How is this to be done? Reporting raw scores will prove little in an experiment of any size since no one is prepared to follow every mathematical operation. Derived scores are little better. The problem is the large amount of data to be handled. The use of one hundred experimental animals results in sixteen hundred separate data, not including peripheral measures such as weight, food intake, preconditioning scores and box or cage numbers. Obviously, the data must be reduced and the only question is how this is to be accomplished.

It might be useful to define exactly what measurement is considered critical. Which variable is the most important in assessing the success or failure of the experiment? Certainly the absolute number of presses and even the mean percent left scores of the groups are of minor importance. I believe the critical factor in our experiments is the measurement of the difference between two corresponding groups. It is not the magnitude of the difference that is important but the interpretation of the difference and the identification of the source of the observed difference. We want to know if the difference is significant. This is, in fact, the quantity or quality that is measured by procedures which determine the statistical significance of a difference. So we find we have come full circle and have deduced that the probability that a difference is real or, conversely, the probability that a difference is not significant, is the critical parameter that we wish to measure.

Careful reflection will show that it is not the use of statistical procedures or the calculation of probability values that is objectionable, but the conversion of the results from quantitative data to quantal data. I find it difficult to believe that a probability of 0.049 is really more significant than a probability of 0.051 and yet in many cases the quantitative difference is ignored and a qualitative distinction is made. A recent article by Pauls (1968) discusses the reasons for maintaining data in a quantitative form and although he did not discuss this particular situation, I believe many of the same arguments are appropriate. I have therefore retained the quantitative characteristics of the data and reported the actual probability values.

3.3. Results

My only purpose here is to orient the reader to the type of investigations performed and the nature of the outcome. Our first series of experiments involved the behavior of the active material in an acetone fractionation procedure. We tested the fractions of an aqueous extract of brain precipitated by various concentrations of acetone and found the active material in the 60–90% fraction. This confirmed the earlier report by Rosenblatt et al. (1967).

TABLE 3.1

Probability values associated with group differences.

	0.025 Dose (N = 102)		Controls (N = 49)	
	Raw scores	Derived	Raw scores	Derived
Average pretest	0.082		0.724	
Test 1	0.024**	0.017**	0.580	0.500
Test 2	0.020**	0.042*	0.871	0.739
Test 3	0.066	0.095	0.560	0.433
Test 4	0.031*	0.040*	0.892	0.764
Test 5	0.078	0.145	0.924	0.869
Test 6	0.017**	0.025**	0.931	0.892
Mean	0.032*	0.038*	0.851	0.730

* Significance using single-tailed probabilities
** Significance using two-tailed probabilities

Our second series of experiments was designed to produce a dose–response relationship for our extract. We found positive effects at 0.025 brains/rat and at about 0.25 brains/rat with an area of ineffective doses in between. This confirmed the triphasic dose–response curve reported earlier by Rosenblatt and even confirmed the actual location of the effective doses. Our final series of experiments examined the reproducibility of our own findings at 0.025 brains/rat. These results are shown in table 3.1. It can be seen that the replication was successful and also that the use of derived scores yielded qualitatively the same results as the use of raw scores.

3.4. *Discussion*

3.4.1. *Validity and significance*

When significant results are mentioned today, the reference is almost always to statistical significance which is usually defined as a probability value less than 0.050. I would not deny the value or importance of statistical methods, but I would like to question the practice of deciding the success or failure of an experiment solely on the basis of a statistical test.

Is it necessary to have a statistically significant difference in order to have a valid effect? I do not believe that it is; in fact there are situations where it is impossible to show statistical significance. Most statistical procedures assess a difference in relation to the variability of the two samples involved. If the actual difference is very small or if the inherent variability is very large, the demonstration of a statistically significant difference becomes extremely difficult. In fact, if the means and standard deviations for two samples are known, there are equations available for the calculation of the probability that statistical significance can be demonstrated.

Certainly, statistical significance is one way in which to demonstrate the validity of a phenomenon, but I believe there is another equally important procedure for achieving the same end. This involves the confirmation of aspects of the phenomenon other than the paramount effect. We have done this in the case of memory transfer and I personally find it very convincing. It is a simple matter to argue that positive transfer effects are not convincing because statistical procedures and derived scores and similar data manipulations confuse the facts, but it is not easy to explain how independent investigators can duplicate results on fractionation procedures and determine almost identical dose–response curves. If the active extracts are not real, how can two investigators find identical behavior on a Sephadex column? This type of replication transcends ordinary statistical procedures and indicates a certain validity regardless of the probability values obtained.

3.4.2. *Replications*

Many objections to the validity or significance of the memory transfer phenomenon have been based on the low success level of attempted replication. What is often overlooked is the fact that most reported replications have a few major variations and many minor variations and therefore cannot properly be called replications at all.

One incident in my own laboratory impressed on me the importance of minor details in experimentation such as memory transfer where all of the important parameters are not known. We had routinely received our recipient animals on Tuesday, deprived them of food on Saturday, and administered the first preconditioning session on Monday. The animals received further preconditioning trials the rest of the week and entered the pretest sessions on the following Monday. After a series of successful experiments, we obtained a set of data which made little sense. The animals scored either between 0 and 10% or between 90 and 100% on every test session, indicating a very strong bias. The reason was a seemingly insignificant change in the routine schedule.

We had become cramped for space and had received our recipient animals on Friday instead of Tuesday. This resulted in the animals entering the preconditioning sessions at a higher motivational level due to a greater degree of deprivation. The animals therefore learned their task faster (cycling in the single lever box) and performed at a higher rate in the initial pretest session. This higher rate of performance in the double lever box led to earlier perseveration and before the injections had any effect, most rats had 'fixed' on one lever. This habit, once acquired, is quite strong and completely masked any effect due to the injection.

The remedy for this situation was quite simple. The preconditioning sessions were put on a criterion basis. The sessions were terminated after 10 min or 25 responses, whichever occurred first. This effectively limited the amount of training a rat could obtain during preconditioning and resulted in a much more homogeneous distribution of scores in the pretest sessions.

It is extremely difficult, if not impossible, to replicate an experiment exactly. Changes are required due to differences in available equipment, manpower, or subjects. The point I would like to make is that these changes can be called minor and insignificant only if the results of the original experiment are replicated. Otherwise, the changes must be suspect. If the changes are major, the replication is not systematic and a failure has no bearing on the original experiment.

Many of the reported replications have indeed contained major changes. The most common is the strain of animal or even the species. To my knowledge, no one has reported the successful transfer of single trial learning and yet reports of unsuccessful attempts keep appearing, claiming to be replications.

3.4.3. Reproducibility

Much has been made of the fact that in the memory transfer field positive results beget positive results. Frequently, the first positive result obtained in a particular laboratory turns out to be the first of a series of positive results. This should not be surprising, however, and is almost predictable. After all, reproducibility is simply the replication of your own work. If replications of the work of other laboratories are difficult, then auto-replications, using the same equipment, procedures and manpower, should be considerably easier.

Even so, conditions will change with time and if a change occurs which affects an important parameter, it can result in the failure of the experiment. Searching out the cause of the failure in this case is usually far more difficult than eliminating it once it has been identified. We have found that the induction of a position habit, our particular brand of memory transfer, is quite reproducible under our conditions. However, we have had two definite failures. One, discussed earlier, was traced to a schedule change and the other was traced to an error in the donor training schedule which resulted in each donor receiving 100 min of training instead of the usual 150 min. In each case, correction of the change led to the immediate reoccurrence of positive results.

3.4.4. Biased data

There are two types of bias which must be considered and dealt with. There is the individual bias of an animal and there is the collective bias of a group. Both are always present, but frequently at a level which does not interfere with the analysis of the results. Extreme individual bias can be eliminated by imposing selection criteria or corrected by using derived scores. Extreme group bias is usually avoided by random assignment of animals to treatment groups but occasionally occurs nonetheless. When present, it too can be corrected by the use of derived scores.

In order to check these points, we performed the analyses shown in tables 3.2–3.5. First (table 3.2), we pooled the data for all of the rats which had received the optimum dose of the extract (0.025 brain/rat). We then calculated the test results with various limits on the acceptable amount of individual bias in the pretest sessions. Although some statistically significant differences persisted, the level of significance was drastically reduced in all cases. However, if one looks at the p value for the average pretest scores, a strong negative group bias is seen to appear as the individual bias is eliminated.

TABLE 3.2

Probability values associated with group differences for pooled experimental data. Statistical analysis was performed on raw scores.

Average pretest limits	Average pretest	Tests 1	2	3	4	5	6	Mean
0–100	0.12	0.004	0.0006	0.002	0.009	0.025	0.007	0.005
20– 80	0.28	0.03	0.003	0.013	0.04	0.09	0.03	0.02
30– 70	0.35	0.063	0.001	0.005	0.014	0.03	0.008	0.008
35– 65	0.81	0.28	0.017	0.07	0.07	0.10	0.029	0.045
40– 60	0.93	0.30	0.024	0.09	0.19	0.14	0.03	0.07

We therefore repeated the analyses but used derived scores instead of actual percentages. These scores were calculated by simply subtracting the average pretest score for each rat from the test score of that rat. In other words, the average pretest score of each rat was used as a baseline for that rat and his deviation from that baseline during the test sessions was measured. The results of these analyses are shown in table 3.3. This procedure resulted in an increased level of significance in those cases where the strong negative group bias was present.

If the foregoing assumptions are valid, then there should be no need for derived scores in the absence of bias. Moreover, analysis of derived scores should correspond to analysis of raw scores. This situation can be seen to exist in the case of the 30–70% group. These limits effectively eliminate

TABLE 3.3

Probability values associated with group differences for pooled experimental data. Statistical analysis performed on derived scores.

Average pretest limits	Average pretest	Tests 1	2	3	4	5	6	Mean
0–100	0.12	0.008	0.0006	0.002	0.005	0.030	0.003	0.001
20– 80	0.28	0.06	0.004	0.008	0.024	0.08	0.01	0.007
30– 70	0.35	0.09	0.0007	0.004	0.01	0.038	0.003	0.003
35– 65	0.81	0.12	0.003	0.019	0.034	0.059	0.008	0.008
40– 60	0.93	0.064	0.004	0.021	0.092	0.075	0.007	0.013

extreme individual bias and there is no strong group bias present. If one compares the p values obtained for this group using raw scores (table 3.3) and derived scores (table 3.2), the similarity is quite remarkable.

As an additional check, the same analyses were performed on the pooled control data. The results are shown in table 3.4 (raw scores) and table 3.5 (derived scores). Although the lack of significant differences makes the interpretation more difficult, the overall trends are similar to those seen in the experimental data. As the extreme individual bias is eliminated, the bias on the test sessions is reduced but a shift occurs in the group bias on the average pretest. The use of derived scores yields a more uniform picture and in the event of little or no bias of any kind (35–65% group) derived scores yield values very similar to those obtained with raw scores.

TABLE 3.4

Probability values associated with group differences for pooled control data. Statistical analysis performed on raw scores.

Average pretest limits	Average pretest	Tests						Mean
		1	2	3	4	5	6	
0 –100	0.94	0.85	0.98	0.74	0.85	0.89	0.89	0.91
20– 80	0.80	0.56	0.91	0.44	0.55	0.76	0.83	0.74
30– 70	0.67	0.68	0.97	0.69	0.76	0.90	0.96	0.90
35– 65	0.49	0.66	0.98	0.79	0.77	0.72	0.89	0.88
40– 60	0.21	0.41	0.97	0.53	0.38	0.31	0.54	0.60

TABLE 3.5

Probability values associated with group differences for pooled control data. Statistical analysis performed on derived scores.

Average pretest limits	Average pretest	Tests						Mean
		1	2	3	4	5	6	
0–100	0.94	0.47	0.82	0.24	0.37	0.50	0.66	0.40
20– 80	0.80	0.31	0.67	0.21	0.27	0.53	0.69	0.37
30– 70	0.67	0.66	0.92	0.46	0.58	0.83	0.91	0.75
35– 65	0.49	0.86	0.98	0.78	0.75	0.77	0.90	0.87
40– 60	0.21	0.77	0.98	0.63	0.45	0.45	0.61	0.35

3.5. Conclusions

Conclusions and generalizations are always dangerous and in this field, with so many unknown and poorly understood parameters, categorical statements would be foolhardy. Nevertheless, conclusions are useful if only to make someone think long enough to disagree and work hard enough to disprove.

What we have in the memory transfer phenomenon is a valid example of behavior induction – the transfer of a tendency which will strengthen or wane depending on the normal reward criteria. Its potential is not in some magic potion for the creation of instant learning but in its use as a tool in understanding the learning process. After all, no one has demonstrated the transfer of conceptual learning. It must be remembered also that learning to press a lever or to perform a discrimination task is a monumental accomplishment for a rat, and yet only the tendency is transferred to the recipient. It is the fact that something can be transferred that is important and not the identity or the extent of the transferred material and behavior.

The problems of the field are enormous. A recent volume devoted to the problems and pitfalls of drug evaluation contains some 450 pages describing the effects of dose forms, dose levels, species, time of observation and many other parameters (Tedeschi and Tedeschi 1968). All of these considerations apply to the evaluation of an extract from donor animals, but in addition, the problems and pitfalls of natural product isolation have to be considered. There are so many places for errors that it seems almost impossible for such an experiment to work.

Nevertheless, the memory transfer phenomenon has been observed and confirmed in several forms. When discussions of the validity of the phenomenon finally replace the discussions of its possibility, then we may begin to realize the full potential of this exciting discovery.

References

BYRNE, W. L., ed., 1970, Molecular approaches to learning and memory. Academic Press, New York.

PAULS, J. F., 1968, *In:* D. H. Tedeschi and R. E. Tedeschi, eds., Importance of fundamental principles in drug evaluation. Raven Press, New York.

ROSENBLATT, F., 1967, Final report on contract no. NONR 401 (59), Studies of memory transfer in rats, AD 656 488, July 1967.

ROSENBLATT, F., 1970a, Induction of behavior by mammalian brain extracts. *In:* G. Ungar, ed., Molecular mechanisms in memory and learning. Plenum Press, New York, London, pp. 103–147.

ROSENBLATT, F., 1970b, Induction of discriminatory behavior by means of brain extracts. *In:* W. L. Byrne, ed., Molecular approaches to learning and memory. Academic Press, New York, pp. 195–242.

SIEGEL, S., 1956, Nonparametric statistics for the behavioral sciences. McGraw-Hill, New York.

TEDESCHI, D. H. and R. E. TEDESCHI, eds., 1968, Importance of fundamental principles in drug evaluation. Raven Press, New York.

CHAPTER 4

Chemical transfer of positively reinforced training schedules: evidence that the effect is due to learning in donors

EJNAR J. FJERDINGSTAD
Institute of General Zoology, University of Copenhagen, Copenhagen, Denmark

4.1. Introduction

Apart from the studies on simpler types of learning such as habituation, the different reports on 'memory transfer' may be divided into two groups according to the type of reinforcement used in training donor animals, i.e. negative or positive reinforcement. It is interesting that of the four original reports in mammals, three, those of Babich et al. (1965), Fjerdingstad et al. (1966) and Reiniš (1965) used positive reinforcement of donors (the fourth, Ungar and Oceguera-Navarro (1965), was concerned with the transfer of habituation). Later studies (Ungar and Irwin 1967; Gay and Raphelson 1967) have shown that negatively reinforced behaviors may equally well be transferred.

There are of course practical advantages in the use of negative reinforcement; e.g. the animals do not have to be deprived and maintained at a certain level of deprivation. For studies of transfer, however, it might be unfortunate that the electric shock used represents a rather severe stress to the donor animal. Thus the objection might be raised that such studies are not able to distinguish between a transfer of learning and one that might be caused by some chemical stress factor. This can be controlled for of course, but is only rarely done.

Studies using donors trained with positive reinforcement schedules, on the other hand, are not as subject to that kind of criticism. Deprivation and subsequent feeding or drinking are certainly treatments which fall within the natural experience of any animal, and it is easier to control for side effects. In the following pages a series of studies all involving positively reinforced donor training, and some of them controlled for the effects of deprivation and reinforcement will be described.

4.2. Procedures and results

4.2.1. Studies with a two-alley runway

In the summer of 1964 the present author and his co-workers carried out a preliminary experiment that furnished the first published indication of the possibility of memory transfer in mammals (Fjerdingstad et al. 1965). Since we were concerned that we should not be deluded by stress related phenomena, we chose to use a positively reinforced type of training; and in order that it should be easy (or so we thought) to do experiments on the specificity of the effect, a two-choice discriminative operant conditioning was decided upon.

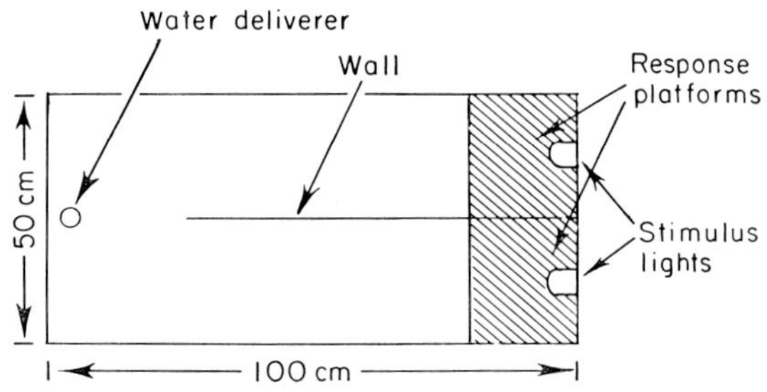

Fig. 4.1. The two-alley runway used in the first experiments. The stimulus lights indicate which platform the rat must step on in order to be reinforced on returning to the water deliverer.

Fig. 4.1 shows a floor plan of the training setup. At the beginning of a donor session a rat deprived of water for 24 hrs was put down at the entrance to the alleys and was faced with the choice between a lit and a dark alley. A run down to press the platform at the end of the lit alley was reinforced with 0.1 ml of water on the rat's returning to the water cup. After consuming the water, the rat would again face the choice between the alleys where the stimulus light was changed in a random fashion. Since it appeared advantageous to keep all subjects equally deprived, the rats were run to a fixed number of reinforcements, 60 per day, rather than to a constant number of trials, and they did not receive water outside the setup. The training equipment was completely automated, so that there was no interference with the animal during a session. Donor rats were run until they reached a criterion of less than 10% errors (percent of reinforcements). Subsequently they were sacrificed with ether, and the brain was removed to be frozen in dry ice and stored in the same material.

From the planarian studies (McConnell 1962) and the results of Hydén and Egyházi (1964) we drew the conclusion that it would appear advantageous to prepare an RNA extract from the trained brains rather than to use a simple homogenate as did some of the other groups which were working independently at the same time. The RNA extraction procedure of Laskov et al. (1959) originally designed for the preparation of RNA from mammalian liver, was slightly modified according to the procedure of Haselkorn (1962). The resulting procedure may be described as a cold phenol extraction in which the RNA after being precipitated once, was purified by redissolving, 18 hrs dialysis against 10^{-3} M NaCl, and renewed precipitation. Later tests showed that although most of the material was RNA, it did contain protein and DNA in amounts corresponding to about 50 and 15% of the RNA content, respectively. This extract was used in the first series of experiments, but was later modified in order to obtain a more pure RNA (see below).

As route of administration to the recipient animals, intracisternal injection was chosen. This was partly because it was supposed that other routes of administration would reduce the likelihood of any effect on the brain, although that has later been proved to be incorrect. Secondly, it has been proven by Brightman (1965) that molecules as large as ferritin (M.W. about 500,000) will penetrate into the cerebral parenchyma when injected in this way.

The actual injection procedure was carried out under anesthesia (at first ether was used, later chloral hydrate). A small incision was made in the skin of the neck, and a no. 27 hypodermic needle attached to a tuberculin syringe was introduced through the foramen magnum into the cisterna magna, i.e. the subdural space immediately behind the cerebellum. The amount of RNA derived from one brain, that is about 1.3 mg, dissolved in 50 μl of a solution of 0.145 M $NaHCO_3$-0.005 M $KHCO_3$ was injected. In some animals the injection was followed by immediate death due to damage to the respiratory part of the medulla following a too deep penetration, but in most cases no ill effects were seen and the animals survived indefinitely. The injection was initially practiced by injecting 1% methylene blue solution.

In the first recipient test three groups of rats were run. The experimental group received extract of donors trained as already described, one control group received a similar extract from the brains of naive donors, and a second control group was not injected but was anesthesized as were the others.

Testing was done exactly like the original training of the donors, except that two short sessions of only 15 reinforcements each were given at 12 and 24 hrs after injection, to be followed by regular sessions with 24 hrs interval. It was observed (fig. 4.2) that the animals of the experimental group for several days

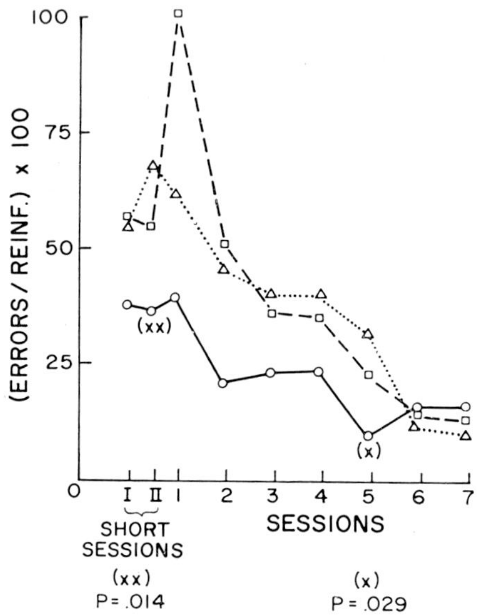

Fig. 4.2. Result of recipient testing in the first experiment. The curves show errors per 100 reinforcements as function of sessions. ○———○ 4 animals injected with trained extract; △........△ 4 animals injected with naive extract; □--------□ 5 animals that were not injected. The group that received the trained extract shows a superior performance from the vary start of testing and through several days. In this and the following figures, unless otherwise stated, p values were calculated by one-tailed U test for the differences between recipients of trained and naive extracts.

had a performance that was significantly superior to that of the two other groups, between which no significant difference was seen. The results thus indicated that RNA extract from untrained animals did not affect the behavior, while extract from trained brains had a definite effect on recipients exposed to the original training situation.

It has been argued that results such as these do not show actual memory transfer, since the test is done under conditions where the desired response is reinforced; the possibility exists therefore that the effect could be due to an enhancement of learning only, an effect which can be caused by a large number of different drugs (see Deutsch 1969). On observing the curves in fig. 4.2, however, it becomes apparent that such an explanation is untenable in this case. If the effect of the trained extract were one of enhancement of learning, the learning curve of the experimental animals should start at about the same level as that of the two other groups and then show a steeper slope. In contrast it actually starts at a much lower level of errors and shows a less steep slope, so that in the end all three groups end up having nearly the same performance. A transfer of the learned tendency in donors to go to the lit alley might be expected to cause just such an effect: since the animals start out making rather few errors because of the transferred preference for the lit alley, they will not be greatly motivated to improve their performance.

As the first experiment seemed to indicate the possibility of a chemical transfer of learned information in mammals, the logical next step was to investigate the specificity of the observed effect. As an approach to this problem two donor groups were trained in the setup in the way already described, except that one group was now reinforced for choosing the dark alley rather than the lit alley. Thus two types of trained extract, 'light-trained' and 'dark-trained', were obtained which were injected into the recipient groups. Two experiments of this type were run; in one both recipients of light-trained extract and recipients of dark-trained extract were tested with reinforcement for choosing the dark alley, in the second experiment the opposite was true, i.e. both types of recipients were reinforced for going to the lit alley. The recipients in both cases were not naive rats, but were trained before injection to run without stimulus lights in the setup until they were able to obtain 60 reinforcements in less than 30 min. Thereafter they were divided into two matched groups and injected.

On testing these recipients it was found that performance was indeed dependent on the type of extract injected, but that the relationship was the opposite of what would have been predicted; recipients of dark-trained extract had the better performance when tested with the lit alley as correct, and recipients of light-trained extract had the better performance when dark was correct (figs. 4.3 and 4.4; Nissen et al. 1966). This was the first example of what has been called a 'reversed effect'; similar results have been reported by other investigators (Rosenblatt and Miller 1966; Ungar and Irwin 1967). Although this is still badly understood, it seems that such reversals are most likely to be obtained in situations where two opposite possibilities of choice,

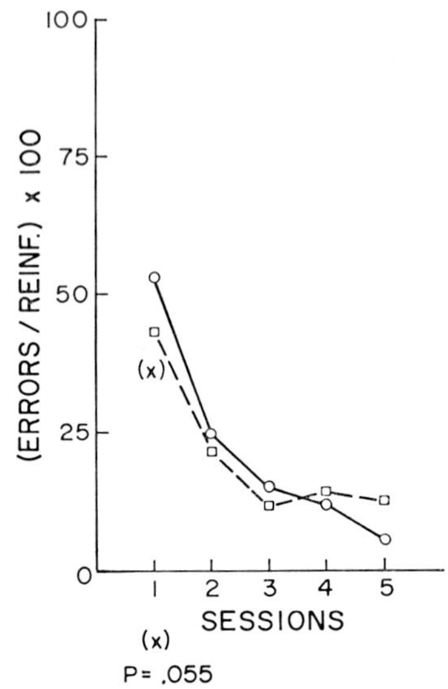

Fig. 4.3. Recipient testing in the first specificity experiment in the two-alley runway. During testing animals were reinforced for going to the dark alley. □--------□ 3 animals injected with light-trained extract; ○——○ 7 animals injected with dark-trained extract. The effect is reversed, i.e. light-trained recipients have a better performance even though training is to go to the dark alley.

such as light–dark or right–left, are used. Also it may be dependent on the dose of extract used (Rosenblatt and Miller 1966).

At the time it appeared that specificity experiments might still be premature, and that it might be advantageous to return to the simpler approach used in experiment I. On doing so it was found that it seemed more difficult to obtain any significant transfer when larger groups were used. It was found that the activity of the trained extract decreased quite consistently with the time expired from dissolving to injecting it (fig. 4.5). Since it seemed possible that this might be due to contamination with hydrolytic enzymes, a method yielding a purer RNA preparation was developed by introducing centrifugations at 25,000 *g* immediately after isolation of the water phase and after redissolving the RNA. In this improved extract it was found that protein contamination had been reduced from about 50 to about 10% of the RNA content.

A series of four experiments was run in which the improved extract was

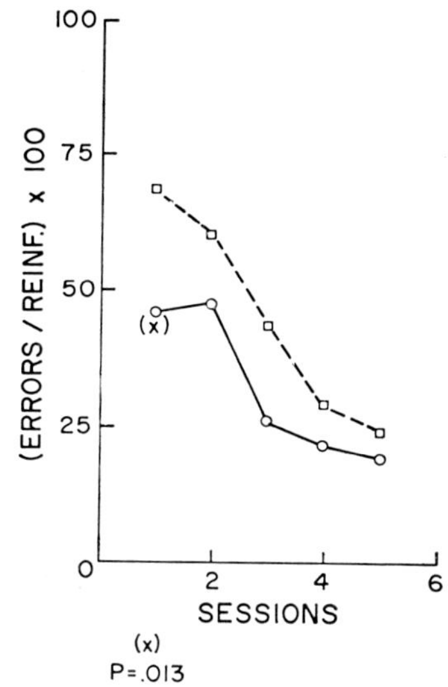

Fig. 4.4. Testing in the second specificity experiment. During the test runs to the lighted alley were reinforced. □--------□ 6 animals injected with light-trained extract; ○———○ 6 animals injected with dark-trained extract. Again the effect was reversed, i.e. dark-trained recipients had the better performance.

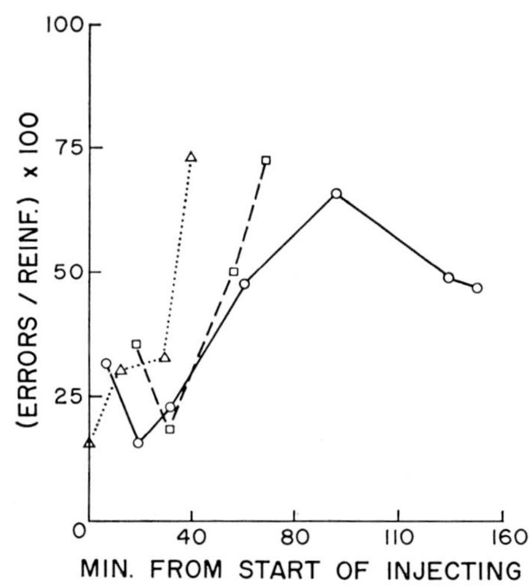

Fig. 4.5. Performance in the two-alley runway of individual recipients in three experiments plotted against the time elapsed from the first animal in the group was injected. The extract is less effective in the animals injected last, indicating instability of the active component. The difference between animals injected before 40 min and those injected later is highly significant ($p < 0.001$).

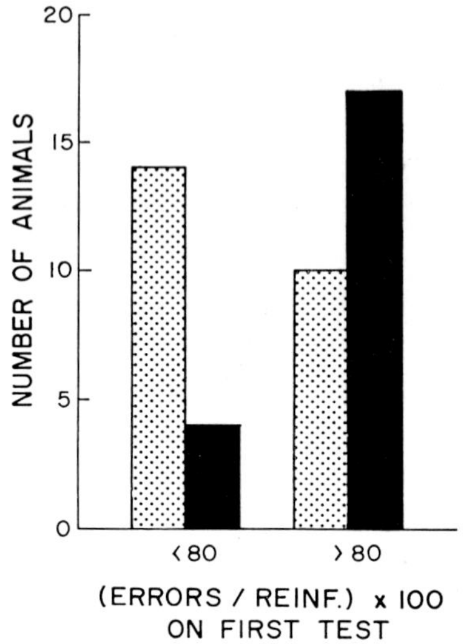

Fig. 4.6. Performance of recipients on the first day of testing in three experiments with an improved technique in the two-alley runway. The effect is reversed, i.e. recipients of trained extract (black columns, n = 21) make more errors than recipients of untrained extract (stippled columns, n = 24). This is highly significant (p = 0.002).

used and the injection procedure changed so as to minimize delay in injecting. At the same time it was decided not simply to inject brain equivalents as this might not be exactly the same amount in each donor group, but rather to measure the nucleic acid content by UV absorption and to inject 1.0 mg per recipient. This corresponds to about $\frac{2}{3}$ to $\frac{3}{4}$ of a brain equivalent.

Very consistent results were obtained when using this approach in a total of 45 recipient rats (Røigaard-Petersen et al. 1968). On the first day after injection there was a significant reversed effect, i.e. recipients of trained extract performed less well than recipients of naive extract (fig. 4.6). On comparing recipient scores with the initial scores of donor animals it was clear that this actually resulted from an increase in preference for the dark alley in recipients of trained extract. At the same time there also seemed to be a transient enhancement of performance in recipients of the naive extract.

This initial reversal disappeared in subsequent tests and was followed by a difference in the predicted direction reaching its maximum after five to six days (fig. 4.7); that is, by then the recipients of the trained extract performed significantly better. The reason for this biphasic effect is not easy to see. There is the possibility that it might be due to a delay in the uptake of the material, which would cause a slow increase in concentration. Alternatively it might be proposed that the injected extract acts only by inducing the synthesis in the recipient of a substance which is the direct cause of the effect. Other psycho-

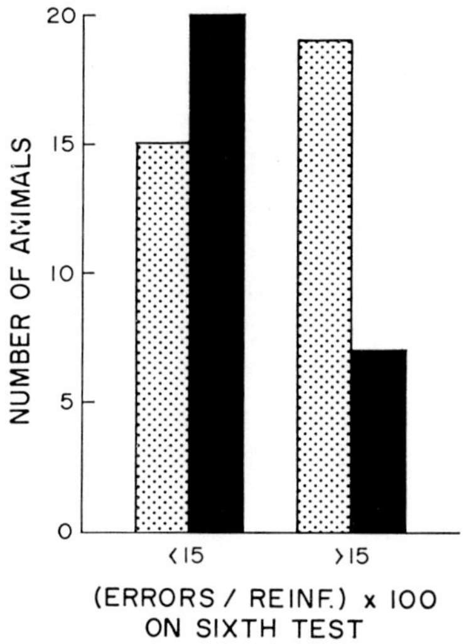

Fig. 4.7. Performance of recipients on the sixth day of testing in the same series of experiments as in fig. 4.6. The recipients of trained extract (black columns, $n = 34$) now have a better performance than recipients of untrained extract (stippled columns, $n = 27$; $p = 0.016$).

pharmacologically active substances, for instance diisopropylfluorophosphate and physostigmine (see review by Deutsch 1969), are known to facilitate performance of a certain task at one time and at another time to inhibit it. Thus the strange reversed effects cannot be considered to disprove the transfer concept as such.

4.2.2. Transfer of right–left discrimination

In the autumn of 1967 the present author joined Dr. William L. Byrne at Duke University to continue his transfer studies. This furnished an opportunity to compare the effects of different approaches to extracting the brains and transferring the extract to the recipient animals. Dr. Byrne had earlier published results obtained with supernatants of crude brain homogenates from rats trained on side discrimination in two-bar Skinner boxes. Donors were reinforced with food for pressing one bar only, one group (R-donors) for pressing the right bar, a second group (L-donors) for pressing the left bar. Initially reinforcement was 1:1 but this was increased gradually during 10 days of training until a 10:1 ratio was obtained. Donors were killed 2 to 4 hrs after the final session and the brains immediately extracted by homogenizing in a 0.01 M tris-0.9% NaCl buffer, pH 7.5, and centrifuging for 20 min at 8000 *g*. The brains were not pooled for extraction, but the amount to be injected into one recipient (one or two brains) was homogenized separately. Immediately after the supernatant was prepared it was injected i.p. into naive rats.

Food-deprived recipients of such injections were screened for preinjection rate of pressing and bar preferences in an unreinforced 10-min session immediately before injection. Fifteen hrs after injection they were given a first unreinforced test of 10 min (UN-1) and after 6 hrs more a second such test (UN-2). On the following several days 20-min test sessions with both bars reinforcing 1:1 (Re 1, 2, 3...) were given once a day. The reinforcement served to keep the rats pressing the bars, but as both bars were reinforcing, supposedly did not interfere with choice of bar.

In several studies of this type (Byrne and Samuel 1966; Byrne and Hughes 1967) it was found that recipients of the trained extracts, R or L, learned to press the bars faster than recipients of naive (N) extracts. This enhancement of learning effect was most pronounced in R-recipients.

In the first attempt to repeat this type of experiment (Fjerdingstad et al. 1970) exactly the same procedure was followed, except for one relatively minor deviation. While the original Skinner boxes had been provided with bars that were quite hard to press (>20 g) the boxes used in the replication had bars that were activated by a press of only 2 g. Also, while earlier side-loading boxes had been used, now completely symmetrical top-loading boxes were used. Whether for this or other reasons a somewhat different result was obtained. No difference appeared in rate of learning, but a tendency was found in both R- and L-recipients to go progressively more to the bar that had been reinforcing during the training of the corresponding donors. This change that only appeared in the reinforced sessions could conveniently be expressed as $\Delta\%$ R, that is the change in the percentage of the presses that were on the right bar. From the first to the second reinforced test the observed changes were highly significant when considering recipients that had been injected with extract from two donor brains (table 4.1). A weaker effect in the same direction was also found in recipients of only one brain equivalent.

In a second experiment (Fjerdingstad et al. 1970) the same behavioral approach was used, but this time the improved RNA extract described earlier

TABLE 4.1

First right–left discrimination experiment (crude extract). Values in two brain recipients of % R test Re-2 versus test Re-1.

L-recipients	R-recipients
	+59.2
	+51.2
	+30.2
	+21.0
	+13.0
+ 2.7	
− 1.0	
	−10.0
−17.5	
−19.4	
−40.4	
−50.1	

$p = 0.002$

was used and injected intracisternally. Again, there was a progressive preference for the bar reinforced during donor training, and even the total range of recipient performance was very similar to that of the preceding experiment (table 4.2). However, it should be noted that since only 1 mg of RNA was injected, or about two thirds of a brain equivalent, the intracisternal RNA procedure meant a large reduction in the number of donors needed. Furthermore, the amount of dry matter injected was 50–100 times smaller in this experiment.

These results then seemed to indicate that under optimal conditions a highly specific transfer of the learned response could be obtained; and secondly that such results could be produced with two widely different types of extract administered by two different routes. Since some other results had

TABLE 4.2

Second right–left discrimination experiment (RNA extract).
Values of % R test Re-2 versus test Re-1.

L-recipients	R-recipients
+58	
	+53
	+26
	+24
	+ 7
− 1	
	− 2
− 9	
−12	
	−12.2
	−13
	−13
−20	
−21	
−28	
−36	
−49	

$p = 0.05$

been published that were interpreted as showing that RNA extracts were without transfer activity, this was an important observation (Luttges et al. 1966).

4.2.3. Transfer of positively reinforced training in the Gay and Raphelson setup

In the fall of 1967 Gay and Raphelson reported very strong transfer effects in a series of experiments in which they used i.p. injected rather crude RNA extracts and at the same time introduced a novel behavioral technique to the transfer field. Donor rats were trained by being put down in the center box S of a series of three connected boxes, from which they were repeatedly pushed into a black box B, and shocked for 5 sec while a gate was closed to make escape impossible. This resulted in a complete reversal of the natural tendency of rodents to stay in the dark, and an increasing tendency to stay in S or go to the box in the opposite side, the white box W. Recipients of extracts from donors trained in this way were tested simply by placing them in S and observing the time they spent in each of the three boxes, with no reinforcement used during testing. A striking reduction in the time spent in B occurred in recipients of trained extract when compared to recipients of untrained extract.

This type of experiment was replicated very successfully by Ungar et al. (1968) who substituted mice for rats as recipients, slightly modified the setup, and injected supernatants of brain homogenates or dialyzates thereof rather than RNA extracts. Also recipients were screened to spend at least 90 sec out of 3 min in the dark box and matched into two identical groups before injection. Although these approaches yielded very striking effects, they might be criticized from the point of view that the apparent transfer could be caused not by learning in donors, but as noted earlier, by a chemical stress factor formed as a consequence of the repeated shocking. Norepinephrine might be a likely candidate. However, it has been shown that injected norepinephrine has no effect on learning and performing a conditioned fear response (Stewart and Brookshire 1969).

Even though it thus did not seem likely that the transfer was caused by the use of shock in negative reinforcement, it still seemed interesting to investigate what effects could be obtained when using the other alternative in donor training, that of reinforcing the rats positively (i.e. rewarding them) for preferring the white box, rather than shocking them for going to the dark box. After joining Dr. Ungar's group at Baylor College of Medicine the author carried out a series of experiments of this type. In all of these the experimental donors, deprived of water for 24 hrs, were trained by putting them down in the center box, facing the lateral wall, and scoring the time elapsed from releasing the rat and until it found 1 ml of water in a small black cup at the far end of the white box. If it did not find the water within a minute, the rat was gently guided there. This was generally necessary only once or twice. Six trials a day were given for 10 days, during which the mean time to find the water decreased from more than 20 sec to less than 2 sec (fig. 4.8). The control donors, rather than being simply naive rats, were deprived to the same extent as were the experimental donors and similarly exposed to the training setup, the only difference being that the water cup was dry. These rats also received 6 ml of water in their home cage before or after being exposed to the setup. Thus in these experiments the effects of the stress of water deprivation and of simple exposure to the experimental environment were also controlled for. Donors were sacrificed 3 to 6 hrs after the last session, and the crude extract described by Ungar et al. (1968) was prepared by homogenizing the brains in distilled water, stirring overnight, centrifuging for 1 hr at 75,000 *g*, and lyophilizing the supernatant. The recipients were mice (20 g Swiss males), screened for preinjection preference as already described, and each received the equivalent of 1.25 g brain, which was redissolved in 0.5 ml of distilled water for injection. Testing was done three times a day for several days, beginning 1 to 2 hrs after injection.

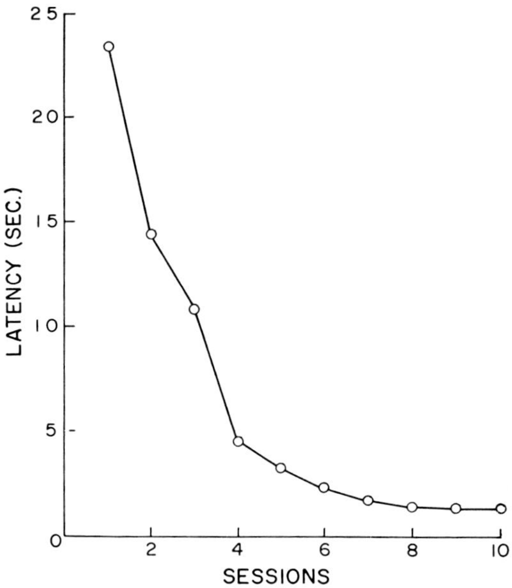

Fig. 4.8. Learning curve of 15 donors reinforced positively for running to the white box in the three interconnected boxes of Gay and Raphelson.

Fig. 4.9 summarizes the results of the six experiments that were run (Fjerdingstad 1969a). Although the effect was less striking than in the experiments using negative reinforcement, there was on the third day of testing a highly significant difference in the predicted direction, i.e. the recipients of extracts from animals that had been rewarded for going to the white box spent less time in the dark box than did the controls.

Thus the change of preference in recipients is not dependent on the type of reinforcement used to induce it in donors. Neither could it be due to effects of deprivation or simple exposure to the experimental environment in trained donors. Therefore the evidence in this case clearly indicates an actual transfer of the learned information as the most likely explanation.

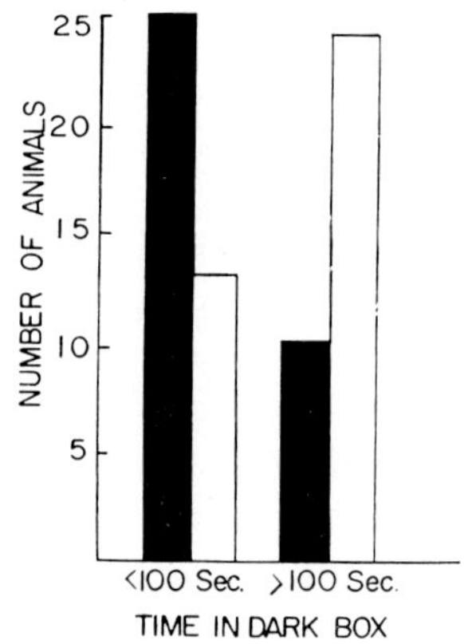

Fig. 4.9. Performance of recipients on the third day after injection in a series of six experiments using experimental donors trained as in fig. 4.8 in the Gay and Raphelson three-box setup. Recipients of trained extract (black columns, n = 35) show a tendency to spend less time in the black box than the controls (white columns, n = 37). This is exactly like the effect that may be obtained with extract from donors trained by repeated shocking in the black box and is highly significant. $P < 0.0025$ (χ^2 test, one-tailed, df = 1).

4.2.4. Transfer of learned alternation

In another series of experiments using Skinner box techniques (Fjerdingstad 1969b) the specificity of the transfer effect was further investigated. As has already been described it had proved possible to transfer right–left discrimination in the Skinner box. Although this at first seems a very good indication of specific transfer, the possibility remains that what is transferred is not a specific 'instruction' to press the right (or the left) bar, but rather some kind of vague preference for one side of the box that only secondarily results in the animal's pressing the bar located in that side. Even though this might seem far-fetched, the available evidence so far had not discounted that possibility. If, however, instead of being trained to press one bar exclusively, rats were to be trained to alternate regularly between bars, and this was found to transfer, any explanation in terms of position preference would be ruled out.

After a pilot study had shown that such transfer was likely to take place, five experiments were run with the same behavioral approach. In each experiment two groups of donors (200 g male Sprague–Dawley rats) were used. Both were deprived of water for 24 hrs before start of training and subsequently received water only in the training setup. In the first five of a series of daily sessions they were reinforced 1 : 1 on both bars, after which they were randomly divided into two groups. One of these, the experimentals, was reinforced only for alternating, i.e. pressing one bar immediately following a press on the opposite bar. This resulted in a rather slow increase in the percent of alternation presses, levelling off at about 75% after three weeks. The other donor group, the control donors, was kept on a schedule where reinforcement was simultaneously available from both bars; a VR-3 schedule was used in order that the duration of a session be about the same in both groups. As might be expected, the control donors soon developed strong preferences for one bar or the other (in more than two thirds of the rats for the left bar), although of course this did not give them any advantage as far as obtaining reinforcement was concerned. Such side preferences are an added difficulty in any type of experiment attempting the transfer of right–left discrimination, since they may also develop in recipients and are not predictable in advance.

Training was carried out for a total of three weeks during which the number of reinforcements (0,1 ml water each) was gradually increased from 60 to 250.

The donors were sacrificed 3½ hrs after the last session. In four of the experiments the crude extract described on page 78 was used and the equivalent of 1.35 g brain injected into the recipients; in the remaining two experiments the improved RNA extract was injected, 0.6 and 0.8 mg respectively. Both types of extract were injected i.p. into 20 g Swiss male mice.

These recipients had been deprived of water for 48 hrs before the injection, and a preinjection test had been given 24 hrs before. During this, as during all subsequent tests, the animals were reinforced 1 : 1 on both bars in a Skinner box fitted with special 'mouse bars'. Each reinforcement was 0.05 ml of water. The performance during preinjection testing was used to eliminate animals with low activity (less than five presses in 20 min) or with too large side preferences (more than 75% on one bar). In the two last experiments animals were also selected to have a low preinjection rate of alternation (less

TABLE 4.3

Transfer of alternation behavior in the Skinner box. Mean number of alternation in recipient groups of experiments II–VI.

Experiment	II		III		IV		V		VI	
	Cont.	Exp.	Cont.	Exp.	Cont.	Exp.	Cont.	Exp.	Cont.	Exp.
Test no.	n = 5	n = 6	n = 7	n = 7	n = 6	n = 6	n = 3	n = 4	n = 7	n = 7
Preinjection	7.67	8.00	7.42	8.14	3.50	2.33	6.83	6.83	3.57	3.86
1	5.75	8.50	8.50	5.50	5.50	4.17	1.67	6.00	3.00	5.60
2	7.75	9.34	7.20	5.80	6.84	5.40	4.50	11.0[1]	2.76	3.75
3	4.20	11.0	7.20	4.14	5.00	7.67	9.00	10.8	3.50	3.80
4	4.67	7.33	5.17	3.14	3.17	6.67[2]	3.00	11.3	5.40	5.40
5	3.00	6.33	4.40	4.20	7.84	5.50	5.50	8.00	6.50	4.60
6	1.60	7.17[3]	5.67	1.75[2]	4.67	3.17	6.34	4.75	6.17	9.34
7	2.20	6.50					6.67	8.25	3.50	7.18[2]
8	2.80	3.84							6.34	9.00
Mean of all test sessions*:	28.00	57.83[4]	28.86	22.00	33.50	31.83	24.33	51.75	29.57	44.83

* Excluding animals that died during testing. In experiment II, one animal of the control group died after 5 days; in experiment III two experimental animals were killed by cage mates after 4 days, and in experiment VI one experimental animal was killed by cage mates after 5 days.

[1] $p = 0.10$; [2] $p < 0.05$; [3] $p < 0.02$; [4] $p < 0.01$

than 33%). The selected mice were matched into two groups with an equal average performance.

After injection the recipients were tested once a day for six to eight days, the first test being given 1 to 2 hrs after injection except for experiment V, where the first test was 24 hrs after injection. The testing equipment was programmed to shut down automatically when a recipient reached 40 reinforcements, or after 20 min had expired.

The results of the five experiments are given in table 4.3. It will be seen that in four out of the five experiments the recipients of the trained extract made consistently more alternations than did the recipients of the control extracts. In one experiment (III) however, there was a slightly higher rate of alternation in the control group, which became significant in the fourth session.

Fig. 4.10 summarizes the results of all five experiments. It may be seen that there is a significantly higher frequency of animals with a high rate of alternation in the experimental group when all recipient animals are compared in this way.

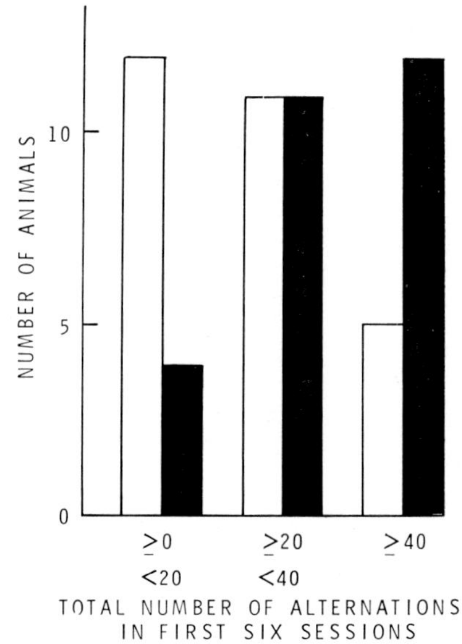

Fig. 4.10. Frequency diagram of the total number of alternations made during the first six test sessions in the experiments on alternation transfer. The experimental animals (black columns, $n = 27$) made more alternations than the control animals (white columns, $n = 28$). This was highly significant. $P = 0.01$ (one-tailed U test), $0.05 > p > 0.02$ (χ^2 test on number of animals in the three categories).

The results, therefore, indicated that alternation behavior in the Skinner box could be transferred as well as could right–left discrimination. Since, in this series of experiments as in the preceding one, the control extracts were derived from donors that had been equally deprived and similarly exposed to the experimental environment and since, as already discussed, the results cannot be due to a position bias or stimulus preference, it would seem that this is a clear-cut indication of specific transfer of learned information.

4.3. Conclusions

The different series of experiments that have been described in the present chapter may be summarized as follows:

The experiments with the two-alley runway demonstrated firstly, that extracts of trained brains were able to influence recipient performance in the corresponding training situation, while extracts of naive brains were without effect. Secondly, later experiments, although yielding the puzzling reversed results, showed that the effect of trained extract was dependent on the type of training applied to the donors. Thus in spite of the reversals some degree of specificity was clearly indicated.

The Skinner box work showed that extract from two groups of donors trained under exactly the same conditions, except for one group learning to press the left bar and the other the right bar, could induce a recipient behavior completely similar to that of the donors. Although this might still be a transfer of side preference bias, at least it would be an instance of specificity.

The experiments with transfer of learned alternation further indicate that what is transferred is actually the learned bar pressing schedule rather than some kind of spatial bias; and these experiments also represent a step forward in complexity of the problem that was transferred.

Finally in the experiments using the three-box setup of Gay and Raphelson (1967) it was found that the transfer of the learned problem was independent of whether negative or positive reinforcement was used to induce learning, and thus could not simply be due to 'fear hormone'-like substances.

The progress in the chemical characterization of the active component of the extracts has been less striking. However, the experiment with right–left transfer definitely demonstrated that the same results might be found with either intracisternal injection of RNA extract or i.p. injection of a rather crude supernatant. The alternation transfer experiments further showed that the RNA extract was also active when introduced by the intraperitoneal route. The main advantage of the intracisternal injection thus remained the reduced amounts of extract needed for transfer.

The similarity of the effects found when using two extracts of such widely divergent composition could be taken to indicate that the active components of both types of extract were identical. Preliminary investigations by Dr. Georges Ungar and the author indicate that this may well be the case (Ungar and Fjerdingstad 1969).

References

BABICH, F. R., A. L. JACOBSON, S. BUBASH and A. JACOBSON, 1965, Science *149*, 656.
BRIGHTMAN, M. W., 1965, Amer. J. Anat. *117/2*, 193.
BYRNE, W. L. and A. HUGHES, 1966, Fed. Proc. *26*, 263.
BYRNE, W. L. and D. SAMUEL, 1966, Science *154*, 418.
DEUTSCH, J. A., 1969, Ann. Rev. Psychol. *20*, 85.
FJERDINGSTAD, E. J., 1969a, Nature *222*, 1079.
FJERDINGSTAD, E. J., 1969b, Scand. J. Psychol. *10*, 220.
FJERDINGSTAD, E. J., W. L. BYRNE, T. NISSEN and H. H. RØIGAARD-PETERSEN, 1970, A comparison of transfer results obtained with two different types of extraction and injection procedure using identical behavioral techniques. *In:* W. L. Byrne, ed., Molecular approaches to learning and memory. Academic Press, New York, pp. 151–170.
FJERDINGSTAD, E. J., T. NISSEN and H. H. RØIGAARD-PETERSEN, 1965, Scand. J. Psychol. *6*, 1.
GAY, R. and A. RAPHELSON, 1967, Psychon. Sci. *8*, 369.
HASELKORN, R., 1962, J. Mol. Biol. *4*, 357.
HYDÉN, H. and E. EGYHÁZI, 1964, Proc. Natl. Acad. Sci. U.S.A. *52*, 1030.
LASKOV, R., E. MARGOLIASH, U. Z. LITTAUER and H. EISENBERG, 1959, Biochem. Biophys. Acta *33*, 247.
LUTTGES, M., T. JOHNSON, C. BUCK, V. HOLLAND and J. MCGAUGH, 1966, Science *151*, 834.
MCCONNELL, J. V., 1962, J. Neuropsychiat. *3*, 42.
NISSEN, T., H. H. RØIGAARD-PETERSEN and E. J. FJERDINGSTAD, 1965, Scand. J. Psychol. *6*, 265.
REINIŠ, S., 1965, Activ. Nerv. Sup. *7*, 167.
RØIGAARD-PETERSEN, H. H., T. NISSEN and E. J. FJERDINGSTAD, 1968, Scand. J. Psychol. *9*, 1.
ROSENBLATT, F., 1970, Induction of discriminatory behavior by injection of brain fractions. *In:* W. L. Byrne, ed., Molecular approaches to learning and memory. Academic Press, New York, pp. 195–242.
ROSENBLATT, F. and R. G. MILLER, 1966, Proc. Natl. Acad. Sci. U.S.A. *56*, 1683.
STEWART, C. N. and H. H. BROOKSHIRE, 1968, Physiol. Behav. *3*, 601.
UNGAR, G. and E. J. FJERDINGSTAD, 1969, Molecular Neurobiol. Bull. *2*, 5.
UNGAR, G., L. GALVAN and R. H. CLARK, 1968, Nature *217*, 1259.
UNGAR, G. and L. N. IRWIN, 1967, Nature *214*, 453.
UNGAR, G. and C. OCEGUERA-NAVARRO, 1965, Nature *207*, 301.

CHAPTER 5

Information specificity in memory transfer

KENNETH P. WEISS

Department of Psychology, King's College, Wilkes-Barre, Pennsylvania

5.1. Introduction

Early research concerning the chemical transfer of information between organisms emphasized procedures which would clearly demonstrate the existence of the phenomenon and which would also be sufficiently robust procedurally to be easily replicable; the recent trend has been to emphasize the identification of the underlying chemistry and mechanism. There appears to be considerably less research activity geared primarily to the determination of the range of specificity which may be usefully transferred. A critical review of the literature, however, indicates the possible development of a reasonable argument for the position that no research to date has clearly demonstrated that anything more than a generalized tendency has, in fact, been transferred.

In advance, it seems unwise to exclude the possibility that many seemingly specific transfer effects may be the result of generalized excitation or depression of somewhat localized cortical areas. Although the validity of this explanation cannot be experimentally rejected at present, it becomes somewhat difficult to employ only this explanation when one considers the wide range of seemingly specific behaviors which have been transferred.

Among the behaviors that have been demonstrated are: discrimination between general environmental stimuli, sound discrimination, left or right responding in a T-maze, preferences for a white as opposed to a black box environment, stimulus light on or off discrimination, tendencies to approach unreinforced food cups, and reversal of the instinctive tendency in goldfish to swim to the surface when oxygen-deprived. Rosenblatt et al. (1966) have concluded: 'A wide variety of tasks, with both positive and negative reinforcement, seem to be susceptible to transfer, and (regardless of any additional generalized activating influence which may be present) the transfer is specific to the learning task'. Although I am inclined to agree with Rosenblatt's general conclusion that the transferred responses seem to be task specific, there has been no direct experimental evidence for this task specificity. In addition, a differentiation between behavioral task specificity and information specificity should be clearly established.

In the relevant literature, behavioral task specificity has typically referred to an animal's responding to a transferred task which was the same as or similar to that task a donor animal was trained on. This type of specificity is epitomized in Ungar's (1966) experiment employing a differential response to sound as opposed to a puff of air. In this experiment, a recipient showed decreased startle only to the stimulus to which the donor had been habituated. Although there is a clear relationship between behavioral task specificity and information specificity, the relevant distinction is that information specificity addresses itself to a consideration of the quality and detail of the transferred information, while task specificity is concerned with an organism's responding to the general characteristics of a specific task.

Analysis of the available literature shows that essentially any behavior which is typically used in the laboratory for studying the acquisition, retention and extinction of learning is amenable to chemical extraction and transfer between organisms. However, as the complexity or subtlety of the behavior increases, the data indicating specificity become extremely difficult to quantify. We should become far more efficient in our ability to extract and transfer information as we learn more about the chemistry and mechanism involved in these phenomena.

Although many investigators have attempted to deal with specificity in information transfer, their studies provide only support for the possibility of specificity. For example, when Ungar (1966) found that habituation to a sound or air blast transferred and appeared to be specific to the donor training, and when he found transfer of left and right choice responses in a Y-maze (Ungar 1967; Ungar and Irwin 1967), it was possible to conclude that these apparently specific behaviors could be the results of selective depression or excitation of localized cortical areas.

The problem is to attempt to determine whether the information and behaviors apparently transferred are analogous to higher level learning phenomena or whether they merely appear to be similar to higher order learning. An examination of the plethora of evidence supporting successful 'specific' inter-organism transfer contradicts the premise that simple excitation or depression of localized cortical sites can explain the results.

With consideration to the above problem, an experiment was devised which could distinguish between transferred behaviors which might, however unlikely, be explainable as generalized phenomena and behaviors which would be analogous to higher order and specific learning.

5.2. The experiment

The rationale employed in this experiment has been developed in detail elsewhere (Weiss 1970). Twenty female 60-day-old Sprague–Dawley albino rats (donors) learned to alternate in their drinking behavior between water bottles on opposite sides of a standard double cage. All animals reached a criterion average performance of 90% correct responses during the 60 days of training.

During training each animal was taken once a day from its home cage (which contained an ad lib. supply of food but no water) and placed in the double-sized training cage for 30 min. Two identical delivery tubes were located in the center of opposite sides and 5 cm from the cage floor. These glass delivery tubes had a circular terminal opening of one-sixteenth of an in. At the end of a 7-sec interval, automatic programming equipment, located remotely, electrified the water in the delivery tube from which the animal had been drinking, punishing further attempts to drink from the same tube. The water remained electrified until the animal drank from the delivery tube on the opposite side, which was not electrified. After a 7-sec interval, this tube was electrified and the other tube de-electrified. This process continued for as long as the animal attempted to obtain water. Approaches to either delivery tube were recorded; a permanent record of the sequence of approaches as well as the number of left and right approaches was taken. It took an average of four days for the animals to acquire this rather difficult alternation behavior. At the end of four days the animals' average performance was approximately 50% correct responding. A correct response was defined as making contact with the water in the delivery tube for up to 7 sec, stopping (either by avoiding the shock or escaping from it) and approaching the opposite tube.

Administration of shock and measurement of the animal's approach behavior were accomplished by an electrode inserted through the rubber stopper alongside the delivery tube in each bottle. For both shock and measurement, an electrical circuit was completed between the animal's tongue and his paws on the galvanized mesh cage floor. A training period of 60 days was used to insure that this alternation would become a major and significant portion of the animal's behavior.

Following the training period, animals were anesthetized with Nembutal (6 mg/100 g body weight) injected i.p., and the entire brain was excised. The brains were immediately frozen in liquid nitrogen and stored at −20 °F.

Forty-eight hrs later a pooled brain homogenate of the 20 donor brains was prepared. A Potter-Elvehjem homogenizer with a teflon pestle was used to homogenize the brains in 75 ml of 0.9% NaCl-0.01 M tris buffer, pH = 7.5. Complete homogenization was judged by dispersal of the tissue on visual inspection. Homogenization took approximately 35 sec. The mixture was then centrifuged for 30 min at 8,000 *g* in a Servall centrifuge at 0 °C. The supernatant and pellet were separated; the supernatant was used for injection.

After storing the supernatant for 48 hrs at −20 °F, approximately 1.5 brain equivalents were injected i.p. into each of 10 approximately 120-day-old female Sprague–Dawley rats, at 30 min intervals. These animals had had normal access to water in their home cages from standard water bottles and glass delivery tubes similar to the ones used in the test cage. After injection, each animal was returned to its home cage with the water bottle removed and allowed to recuperate for 8 hrs. After this interval each of the 10 animals was sequentially placed in the test cage for a 20-min trial interval. In the test cage the animal had free choice of either delivery tube; no shock was ever administered. The sequence of approach responses was recorded. An approach was defined as making electrical contact with either delivery tube after it had been abandoned for at least 3 sec. At the end of the 30 min test, the animal was removed to its home cage where it remained for $4\frac{1}{2}$ hrs (without water) after which time it was again placed in the test cage for another 30 min test. Animals were provided free access to food in the home cage. Measurements of these 10 animals continued sequentially for five days around the clock with no interruptions.

A control donor group of 20 animals was treated exactly as were the 20 experimental donors except that these animals were given 60 days of irrelevant training in a Lehigh Valley Electronics operant conditioning cage (no. 1578–1642) where they learned, in daily sessions of 30 min, to press a bar for a drop of water on an automated dipper. During the control trials, these animals received no water in their home cages and learned to obtain their water in the sham test cage. The brain homogenates (prepared as for experimental donors) from these control donors were injected i.p. into 10 naive control recipients, who were tested for alternation behavior in the experimental test cage in the same manner as the experimental animals.

5.3. Results

Previous published research (Weiss 1970) and pilot studies conducted in my laboratory have indicated that the maximum effect produced through injection of the 'trained' homogenate with my procedure occurs at approximately 24 hrs after injection, remaining high up to approximately 48 hrs and generally showing significant decrements between 48 hrs and 72 hrs. Prior to this experiment, therefore, it was decided that the critical comparisons for the experimental and control groups would be the sequence of approaches (percent correct approaches) during the first 48 hrs. A correct response again consisted of a simple alternation. The results of the statistical analysis are summarized in table 5.1. Tables 5.2 and 5.3 are percent correct alternation responses for the experimental and donor groups respectively over the first nine test sessions.

TABLE 5.1

Mean number of correct responses (alternations) for each animal across the nine critical trials.

	Experimental group	Control group
	35.0	20.5
	25.5	23.0
	36.3	17.3
	29.1	16.7
	27.3	26.8
	35.1	15.6
	35.7	15.4
	27.5	19.9
	21.8	16.7
	20.4	17.0
Mean	29.37	18.89

$df = 18$; $4.77 > 3.92$; $p_2 < 0.001$ (two-tailed T test)

TABLE 5.2

Percent correct alternations (experimental group).

Trials	Animal designation										Mean across animals
	D	E	F	G	O	C	P	H	R	Q	
1	29	34	40	29	38	33	19	30	3	16	27.1
2	39	31	42	27	16	40	30	27	20	14	28.6
3	33	25	36	25	14	33	36	34	23	17	27.6
4	37	21	41	26	34	45	35	39	28	23	32.9
5	30	27	28	32	33	38	39	35	38	18	31.8
6	44	16	31	34	39	32	44	19	26	24	30.9
7	37	19	39	31	30	34	34	23	31	31	30.9
8	18	28	42	30	18	31	47	17	16	22	26.9
9	48	22	28	28	24	30	38	22	12	19	27.8
Mean across trials	35.00	25.55	36.33	29.11	27.33	35.11	35.77	27.55	21.88	20.44	

TABLE 5.3

Percent correct alternations (control group).

Trials	Animal designations										Mean across animals
	I	B	N	J	L	K	T	V	W	Y	
1	39	32	17	24	100	11	11	25	16	7	28.2
2	25	19	15	4	33	35	25	15	13	19	19.3
3	11	15	13	15	11	11	16	21	24	18	15.7
4	16	25	36	2	14	13	13	15	24	25	19.5
5	25	20	23	27	15	28	25	19	15	25	22.2
6	14	21	17	17	21	9	9	17	19	12	15.6
7	29	31	4	21	21	5	5	23	17	17	17.3
8	17	27	15	29	11	13	13	26	13	10	17.4
9	9	19	16	11	15	25	24	18	9	20	16.6
Mean across trials	20.55	23.22	17.33	16.67	26.77	16.67	15.67	19.88	16.67	17.00	

The behavior of recipients was recorded continuously for an additional 72 hrs in an attempt to obtain information regarding the shape of the 'extinction' curve. Fig. 5.1 is a graph of the mean correct responses (alternations) of the 10 animals in each group during this period.

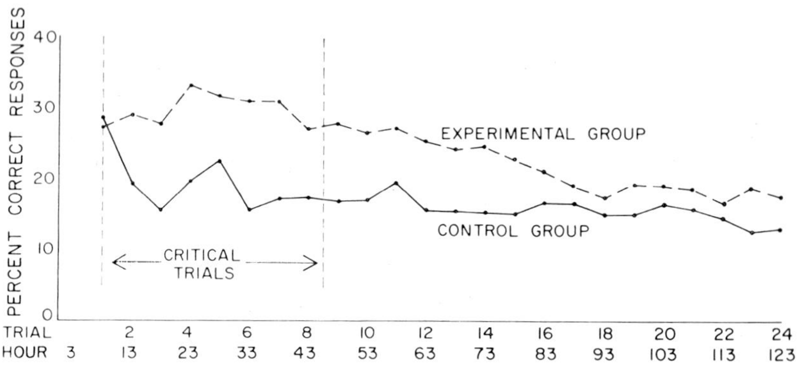

Fig. 5.1. Mean correct responses (alternations) of the ten animals in each group for the five-day (post injection) measurement period.

It is clear from this graph that there is a distinct decrease in the recipient animals' average alternation behavior. The graph, however, does not appear to be a typical extinction curve. It is probable that the animals' alternation behavior would not go below the apparent plateau established at about 90 hrs. The high degree of variability of both the control and naive animals requires careful interpretation of this graph. Many naive animals produce very few alternation responses and develop a relatively strong preference for drinking at one side of the cage. The majority of animals which we have measured on this response, however, exhibit considerable random alternation. For this reason, a graph of the average perfoımance on this task is a relatively poor description of the actual observed behavior. These factors were taken into account in formulating the operational definition of the correct response.

Some normal animals randomly approach either delivery tube approximately 50% of the time. The actual percent of alternations remains equally low for those animals that develop a preference for one side and for those animals who show short-duration preferences only during random sequences of responding. Preliminary experiments indicated, for example, that immediately after an animal awakens, it may spend several minutes drinking from one side of the cage, exhibit grooming and exploratory behavior, and spend several minutes drinking at the other side of the cage (not necessarily continuously on either side). Such an animal may show approximately 50% approaches toward both tubes. However, the number of alternations tends to be extremely low. Thus, while careful records of the percent response of each animal to each delivery tube were maintained, these data are not directly relevant to our present discussion. Only the percent correct responses (alternations) is considered in comparing experimental and control groups.

The methodology and procedure employed in this experiment, it should be mentioned, are the result of four unsuccessful attempts which had to be abandoned before completion due to considerable and unexpected difficulties in developing this paradigm. Considerable difficulty was experienced while simply attempting to establish alternation behavior in donors. The amount of time the donor is allowed to drink from one delivery tube appears to be critical. Drinking periods in excess of 7 sec tend to reduce the likelihood of the animal's alternation to the other tube, and also tended to make it more difficult for the animal to avoid the shock by anticipating the end of the interval. Very short intervals (3 to 4 sec), appeared to frustrate the animals and to generate a high level of aggressive behavior such as biting and clawing at the front of the cage. It was found that a very high shock intensity (limited to approximately 250 mA)* had to be maintained during training. Animals just learning to avoid shock and to alternate when occasionally shocked, would demonstrate considerable aggressive behavior and would regress to non-alternation behavior. These animals would attempt to obtain water from the same delivery tube while enduring considerable momentary shock as they very quickly licked the drop of water at the tip of the tube.

The difficulty and complexity of this particular alternation task are such that all donor animals, although reaching the criterion number of correct responses, developed several different strategies for coping with the experimental situation. Seven of the donors appear to have learned primarily to avoid shock by drinking for only 4 to 6 sec and then alternating. Ten donors

* Although the maximum shock, limited by resistors, was 250 mA, the actual shock received was considerably less, varying within and between subjects, as is typical with shock training. In addition, in this procedure animals would make a very quick pass at the tube with a paw or tongue and not really come into good contact with either the tube or grid floor.

responded almost exclusively by escaping after shock onset. Three animals exhibited random escape and avoidance. Several animals learned to swat the delivery tube with their paws before approaching to drink in an apparent attempt to 'check' their choice.

Both the swatting behavior and the behavior of jumping away from the delivery tube (as if shocked) were observed in several recipient animals.

One possible way to estimate the task difficulty is to analyze the behaviors developed to obtain water without alternation. In one preliminary experiment during which food was available in the test cage, several animals learned to pile food blocks in front of the delivery tube and to avoid shock by standing on these blocks which insulated them from the grid floor. One animal managed to short-circuit both the drinkometer and the shock apparatus by arranging the food blocks in a heap in front of the delivery tube where they became saturated conductors. Several animals learned to swat the end of the tube with their paws and then to lick their paws for water. This behavior was learned apparently because of the less painful shock received through the calloused paw. When shorter time intervals than 7 sec were used, a few animals learned to swat one delivery tube with their paw and then to alternate and drink from the other. Throughout all of the preliminary experiments, it was noted that most donor animals would develop rather specific and apparently strong habits, superstitions and strategies. In the early stages of learning the alternation, all animals emitted a high level of activity.

Two of the donors in the present experiment developed a habit of attempting a 'dry run'. They would take a position approximately halfway between the two delivery tubes; each animal would look at one delivery tube, slowly move its head until it was looking at the other, and then snap its head back to face the first tube. Each of these animals would do this several times before approaching (usually correctly) a delivery tube. This behavior may be related to some kinesthetic feedback (possibly related to the memory of the previous response) which facilitated their choice of correct alternation.

The relevance of these subjective observations to our present discussion is the fact that several of the recipients demonstrated similar behaviors subsequent to their injection. Several independent observers identified several other subtle behaviors and superstitions in recipients that were also observed (casually) in the donors.

The subjective evidence for the validity of these observations has led us to initiate the development of procedures to evaluate and quantify them. We are currently using video tape to record the behavior of the donors and then the behavior of each recipient, employing a one-to-one injection procedure instead of the pooled extract used in this experiment. We hope to be able to study the various collateral behaviors transferred along with the desired response; the relative difficulty involved in learning the task in conjunction with the high level of motivation the task maintains, appears to produce and maintain these individual collateral behaviors.

In summary, the results of this experiment provide substantial support for the thesis that specific and useful information can be transferred between organisms (at least at the level of the behavioral specificity required for alternation). The critical distinctions between this experiment and previous studies are: (1) Recipients were not provided with any external stimulus or cue during their performance of the apparently transferred behavior. (2) The alternation response is incompatible with, and to some extent precludes, an explanation of the transferred behavior as simple differential excitation or depression of specific cortical areas. (3) Information transfer, as opposed to the transfer of facilitation of learning, was accomplished by not introducing a relearning situation through the use of reinforcement in recipient animals. No shock was used on the recipients. (4) The relative degree of task difficulty and complexity which we achieved has not, to our knowledge, been demonstrated previously.

References

BYRNE, W. L., ed., 1970, Molecular approaches to learning and memory. New York, Academic Press.

FJERDINGSTAD, E. J., T. NISSEN and H. H. RØIGAARD-PETERSEN, 1965, Scand. J. Psychol. *6*, 1.

GAY, R. and A. RAPHELSON, 1967, Psychon. Sci. *8*, 369.

GUROWITZ, E. M., 1969, The molecular basis of memory. New Jersey, Prentice Hall.

JACOBSON, A. L., F. R. BABICH, S. BUBASH and A. JACOBSON, 1966, Psychon. Sci. *4*, 3.

LUTTGES, J., T. JOHNSON, C. BUCK and J. MCGAUGH, 1966, Science *151*, 834.

ROSENBLATT, F., J. T. FARROW and W. F. HERBLIN, 1966, Nature *209*, 46.

ROSENBLATT, F., J. T. FARROW and S. RHINE, 1966, Proc. Natl. Acad. Sci. U.S.A. *55*, 548 and 787.

ROSENBLATT, F. and R. G. MILLER, 1966, Proc. Natl. Acad. Sci. U.S.A. *56*, 1423.

UNGAR, G., 1966, Fed. Proc. *25*, 207.

UNGAR, G. and L. N. IRWIN, 1967, Nature *214*, 453.

WEISS, K. P., 1970. Measurement of the effect of brain extract on interorganism transfer. *In:* W. L. Byrne, ed., Molecular approaches to learning and memory. New York, Academic Press, pp. 325–334.

CHAPTER 6

Factors controlling interanimal transfer effects*

WILLIAM B. RUCKER**

Department of Psychology, The George Washington University

6.1. Introduction

Interanimal transfer of training is defined by the influence of donor training on recipient performance. The recipients are not obligated to mimic their respective donors; the only operational requirement is a lawful relationship between the parameters of donor training and the parameters of recipient performance, however contrarily the recipients may behave. In this regard, interanimal transfer of training is similar to the more orthodox transfer of learning defined by the effects of practice of one task on the performance – by the same subjects – of some subsequent task. These effects may be either facilitation of performance of the second task (positive transfer), interference with performance of the second task (negative transfer), or no effect on performance of the second task (zero transfer or a cancellation of equally strong positive and negative transfer effects). Under various conditions, the behavior of recipients in interanimal transfer of training experiments has been observed to be similar to that acquired by donors (positive transfer), opposite to that acquired by donors (negative transfer), or apparently unaffected by training given to donors (zero transfer). Interanimal transfer of training should therefore be considered as a vector characterized by both magnitude and direction. Whereas a score of variables have been found to influence transfer of learning (Ellis 1965), we presently have little

* This investigation was supported by Biomedical Sciences Support Grant FR 07019 (–03 and –04) from the General Research Support Branch, Division of Research Resources, Bureau of Health Professions Education and Manpower Training, National Institutes of Health.

** Present address: Department of Psychology, Mankato State College, Mankato, Minnesota, 56001.

knowledge or control of the factors controlling interanimal transfer of training.

Whereas the study of transfer of learning has been directed towards those factors increasing the generality of transfer, the study of interanimal transfer of training has been directed towards the specificity of the transfer. Consequently, the repeated reports of negative interanimal transfer effects have had no theoretical framework to deal with such apparent failures of specificity. In the earliest report of negative transfer, for example, Fjerdingstad and his associates found that performance of recipients on a light–dark discrimination was improved by injections of an RNA extract prepared from donors trained to the same task, but even greater improvement of recipient performance was obtained with extracts prepared from donors trained on the opposite discrimination (Nissen et al. 1965; Fjerdingstad et al. 1965). The only specificity shown is for the negative component of the transfer effects.

In puzzling over the results reported by Fjerdingstad and his coworkers, Ward Halstead and I felt that a key factor was the great amount of overtraining the Danish workers had given their donor animals. Overtraining is one of the factors contributing to transfer of learning (Ellis 1965). In rats, overtraining on a reasonably difficult discrimination for relatively large rewards facilitates acquisition of the reversed discrimination (Mackintosh 1969). Undertraining tends to interfere with acquisition of the reversed discrimination. This overtraining reversal effect may be viewed as the result of differences between what is learned early in training and what is learned late. It was possible that two learning factors were potentially available for interanimal transfer in the Danish experiments and that these factors could be separated by varying the amount of training given to donors. We predicted that recipient performance on the opposite visual discrimination in our experiment would improve monotonically from negative transfer at the lowest levels of donor training to positive transfer at the highest levels. Instead, we found that recipient performance was a wiggly function of the amount of donor training with: baseline transfer produced by extracts from donors given no discrimination training, negative transfer added as a result of minimal training, positive transfer at moderate levels of training and finally negative transfer with overtraining. These effects were statistically reliable. On the basis of other experimental evidence involving comparisons of two methods of storing donor brains, two methods of RNA extraction and two methods of testing recipients, we nevertheless stuck to the basic idea that chemically distinct transfer agents develop at different stages of donor training (Halstead and Rucker 1970; Rucker and Halstead 1970). Using a proteinaceous extract and a position discrimination, Ungar and Irwin (1967) reported that three-day increments of donor training up to 18 days past original criterion performance produced a steady drop from positive transfer at the lowest levels of donor training to negative transfer at the highest levels of donor training. This result, which runs just opposite to the prediction we had made for our experiment, implies factors other than amount of donor training contribute to negative transfer.

A strong non-monotonic relationship between dose and the direction of interanimal transfer has been reported (Rosenblatt and Miller 1966; Rosenblatt 1970a; chapter 3, this volume). Although it has been difficult to control variability across experiments (Rosenblatt 1969; Rosenblatt 1970b), it appeared in early reports that a dose of proteinaceous extract equivalent to 0.025 parts brain consistently produced positive transfer while half or twice that dose consistently produced negative transfer. This would suggest that active components of the extract differ in dose–response curves and/or that additional receptors in the recipient are brought into play at higher dose levels. Only one active transfer agent has been even partially characterized (a 15-amino-acid peptide under study by Georges Ungar, personal communication). We do not know how many transfer agents may be present in a more or less refined extract, or what the real dose levels may be at any particular dilution of the extract.

Transfer agents might also be expected to differ in their time course of action. It has become customary to test recipients several times on the same task and to analyze each session independently. Transfer results generally differ from test session to test session with 'best' results sometimes reported for the second and third day. A slightly different approach was used by Rucker and Halstead (1970), who first tested recipients without reward and then, as a switch, tested the same recipients on a learning task. Different transfer activities were reported for the two sessions, but the differences could have been due either to the change in task, to the order of the two tasks, or to the time between injection and testing on the two tasks. I have found only one report of independent groups of recipients tested at different times after injection (Schad et al. 1969) and none which compares repeated measures with independent measures.

Golub et al. (1970) report that the positive transfer of a bar-press response can be greatly facilitated by properly scheduled periods of rest between certain phases of donor training. These authors suggest that the enhanced transfer effect is due to superior donor learning, but it may be that the rest periods act to stabilize the transfer agents thus allowing them to act more evenly on recipient performance. Nevertheless, it is going to be very difficult in the future to rule out the possibility that the transfer effects observed under different conditions of donor training are due to differences in the real dose levels in effect at the time recipients are tested. It is even possible that all of the mystery of transfer of training could resolve parsimoniously into such a single factor. In the meantime, it is necessary that we work our way through the many variables that contribute to the dose factor, such as the method of extraction (including storage), the apparent dose administered, the route of injection, the interval between training and sacrifice of donors, the interval between injection and testing of recipients, and characteristics of recipients such as strain, age, sex and deprivation.

In our experiment with different levels of donor training (Rucker and Halstead 1970) the wiggly transfer effects produced may have been due to differences in the real dose in effect at the time the recipients were tested. This is possible (despite our attempts to control the apparent dose administered and despite circumstantial evidence that several transfer agents were involved) because the availability of transfer agents may depend on the amount of training given donors. Our wiggly transfer curve looks suspiciously like Rosenblatt's wiggly dose–response curve.

The experiment reported here simply asks what happens when both the administered dose and the amount of training given donors are varied systematically in a factorial design.

6.2. Methods

6.2.1 Subjects

A total of 104 male rats of the Wistar-Furth/MAI strain were delivered from Microbiological Associates in eight weekly lots of 13 animals each. The subjects weighed 175 to 200 g on delivery. The rats in each delivery were assigned randomly to donor or recipient treatments according to the experimental design described below. The subjects were housed individually throughout their participation in the experiment.

6.2.2. Apparatus

The discrimination apparatus used in this experiment, previously described by Rucker and Halstead (1970), consisted of a start box, a running area, and three goal boxes. When a guillotine door between the start box and the running area was raised, a subject was permitted to descend from the start box into the running area by way of a ramp. The ramp inclined at 20° to allow the natural downward focus of the rat to encompass the three stimulus doors leading to the goal boxes. These doors were framed by 2 in columns so that a response could be counted when the subject got his ears between two columns.

6.2.3. Pretraining

On the three days prior to pretraining, the rats were weighed and handled for two 5-min periods. They were fed 10 g of Purina rat chow each night throughout their participation in the experiment. On handling days, they received an additional 10 Noyes pellets (45 mg, formula A) as preparation for the subsequent use of these pellets as reward. On the three days following handling, the subjects went through three phases of pretraining in the discrimination apparatus. Each phase consisted of six complete trials from start box to a goal box. On each trial the subjects were allowed to eat two Noyes pellets from a trough in the goal box of their choice.

In the first phase of pretraining, all three stimulus doors were left open. In the second phase, the doors were partially closed; in the final phase, the doors were closed. The doors were never latched during pretraining and no discriminative stimuli were used until discrimination training began.

6.2.4. Discrimination training

On each trial of discrimination training, only one of the three doors could be knocked down to achieve the reward pellets. The other two doors were closed and latched. To provide light or dark discriminative cues, a 7.5 W bulb at the back of a goal box could be turned on to shine through the translucent door to the goal box or the bulb could be turned off with a black plastic card placed on the door to block stray light. Donors were trained with light as the positive stimulus and recipients were trained with dark as the positive stimulus. In either type of training, the positions of the single positive cue and two negative cues varied from trial to trial according to a prearranged sequence. The sequence was arranged so as to discourage possible position habits without losing the statistical appearance of randomness. Subjects were never allowed to correct an incorrect response.

A day's training consisted of 36 trials divided into blocks of nine trials. Within a block of nine trials the intertrial interval was controlled by the time it took to replace a subject in the start box and to change the stimulus cues according to the sequence schedule. This took approximately 30 sec. The intersession interval between blocks of trials allowed all of the subjects in the donor or recipient group to complete that block. For the recipient group the intersession interval was approximately 90 min.

6.2.5. Experimental design

Rats in each shipment were divided randomly into a group of three donors and ten recipients. The donors were all pretrained and then given either zero, 36, or 108 trials with light as the positive stimulus. Twenty-four hrs after the end of the training, the donors were sacrificed and their brains removed for light homogenization in ice cold saline made up to 6 ml. Five ml of the homogenate were taken up for the high dose preparation (0.835 parts brain) in a saline wetted syringe which was stored on ice until the other doses had been prepared. The homogenizer was washed with 5 ml saline and the wash used to dilute the remaining 1 ml aliquot of homogenate. Again 5 ml were taken as the middle dose preparation (0.139 parts brain) and the remainder diluted with 5 ml saline from which the low dose preparation (0.023 parts brain) was taken in 5 ml. The order of sacrifice of the three donors was randomized as was the order of injection of doses prepared from a given donor brain. Recipients were injected i.p. (20 gauge needle) immediately after preparation of all three doses. In addition to the nine recipients receiving one of three doses from one of three donor sources, a tenth recipient was injected with 5 ml cold saline.

The recipients were injected on the day after they had completed pretraining. Twenty hrs after injection they began discrimination training with dark as the positive stimulus. The order in which they were trained was randomized for each shipment block but remained the same for a given block on each of the three days of training. The experimenters who trained the animals did not know how treatments had been assigned to subjects and the data from the entire experiment was decoded at the end of the project.*

In summary, the design of the experiment called for varying the amount of training given donors over three levels and for varying the dose administered over three levels. Members of the 3 × 3 factorial design and the control group receiving saline injection were tested in eight randomized blocks over three days of repeated measurements. Table 6.1 shows the flow of donors and recipients through the course of their participation in the experiment.

TABLE 6.1

Flow of subjects through one experimental block.

Day	Donors	Recipients
Thursday	Delivered, housed individually, fed ad lib.	Delivered, housed individually, fed ad lib.
Friday	Handled 2 × 5 min, fed 10 g and 10 pellets	Fed ad lib.
Saturday	Handled 2 × 5 min, fed 10 g and 10 pellets	Fed ad lib.
Sunday	Fed 10 g and 10 pellets	Fed ad lib.
Monday	Pretrained – phase 1, fed 10 g	Handled 2 × 5 min, fed 10 g and 10 pellets
Tuesday	Pretrained – phase 2, fed 10 g	Handled 2 × 5 min, fed 10 g and 10 pellets
Wednesday	Pretrained – phase 3, fed 10 g	Fed 10 g and 10 pellets
Thursday	Trained, trials 1–36, fed 10 g and make-up pellets	Pretrained – phase 1, fed 10 g
Friday	Trained, trials 37–72, fed 10 g and make-up pellets	Pretrained – phase 2, fed 10 g
Saturday	Trained, trials 73–108, fed 10 g and make-up pellets	Pretrained – phase 3, fed 10 g
Sunday	Sacrificed	Injected, fed 10 g
Monday		Tested, trials 1–36, fed 10 g
Tuesday		Tested, trials 37–72, fed 10 g
Wednesday		Tested, trials 73–108, scrubbed

* The author gratefully acknowledges the assistance of the following students who were involved either in preliminary work or in the conduct of this experiment: Steven Barker, Arthur Eisenman, Edward Faine, Ronald Holtzman, Terrence Rosen, Steven Silver, Irene Tritter, Catherine Walters.

6.3. Results

Table 6.2 shows the daily performance of recipients in terms of the mean number of correct responses made during each day's training session of 36 trials. Analysis of variance of the factorial part of the experiment (i.e. without the saline group) did not reveal significant main effects due to the level of donor training or due to dose, and did not reveal a significant interaction between donor training and dose. Looking at daily performance for these subjects, the learning curve was quite reliable ($F=587$; $df=2, 170$; $p<0.001$). The performance as a group improved from a mean of 15.0 correct on the first day, 21.8 correct on the second day, to 30.0 correct on the third day. While neither the level of donor training nor the dose administered interacted significantly with the day of testing, the triple interaction between level of donor training, dose and the day of testing was significant ($F=2.73$; $df=8, 170$; $p<0.01$). The interaction between randomized blocks and day of testing was mildly significant ($F=1.80$; $df=14, 170$; $p<0.05$).

In order to make daily performance comparable between groups, and to remove both the effect of learning and daily differences between blocks of replications, the daily scores were converted by computer to z-scores based

TABLE 6.2

Mean correct responses of recipients.

Donor training	Day	Dose			Saline
		low	middle	high	
Pretrained	1	16.9	15.3	16.1	
	2	25.6	20.0	21.0	
	3	32.9	31.6	29.3	
36 Trials	1	16.9	15.4	11.9	
	2	23.0	23.7	19.6	
	3	31.5	31.0	28.4	
108 Trials	1	15.0	14.1	13.5	
	2	18.4	22.6	22.0	
	3	29.6	27.7	28.0	
Saline control	1				14.6
	2				19.5
	3				30.6

on the performance of subjects of a given block on a particular day. Except that the effect of learning and the daily differences between blocks disappear, the analysis of variance produced the same result. Only the triple interaction between level of donor training, dose and day of testing was significant ($F=2.63$; $df=8, 170$; $p<0.01$).

Because of the triple interaction, the results of the experiment must be examined at the level of a single day's performance for each combination of dose and level of donor training. Table 6.3 shows the totals of transformed scores for each group on each day. Any two of these totals may be compared, by Sheffé's method of multiple contrasts (Brownlee 1965), by subtracting one score from the other. Should the absolute value of this difference exceed 3.52, 4.22 or 5.57, then the difference is significant at the $p<0.10$, $p<0.05$ or $p<0.01$ level, respectively.

TABLE 6.3

Totals of transformed scores of recipients.*

Donor training	Day	Dose			Saline
		low	middle	high	
Pretrained	1	3.15	0.80	0.33	
	2	5.80	−2.87	−1.72	
	3	5.67	1.37	−1.86	
36 Trials	1	3.01	1.01	−5.51	
	2	2.77	2.79	−2.03	
	3	1.31	2.26	−1.05	
108 Trials	1	0.29	−0.12	−2.11	
	2	−4.20	1.52	−0.72	
	3	−0.58	−3.38	−3.82	
Saline control	1				−0.83
	2				−2.77
	3				0.06

* Z-scores based on daily performance within randomized blocks

While a large number of contrasts may be constructed on the basis of table 6.3, the effect of injecting brain material from donors not given discriminative training and the effects of increasing levels of donor training are of primary concern. Consequently for each day and dose combination, contrasts were set up to test for the following effects:

Effect A, defined as the difference in performance between recipients of homogenate prepared from donors given no discrimination training and recipients of saline.

Effect B, defined as the difference in performance between recipients of homogenate from donors given 36 trials of discrimination training and recipients of homogenate from donors given zero trials of discrimination training.

Effect C, defined as the difference in performance between recipients of homogenate from donors given 108 trials of discrimination training and recipients of homogenate from donors given 36 trials of discrimination training.

The results of these comparisons are reported in table 6.4.

TABLE 6.4

Three transfer effects: direction and level of significance.*

Effect	Day	Dose		
		low	middle	high
A	1	+	0	0
	2	+ + +	0	0
	3	+ +	0	0
B	1	0	0	− − −
	2	0	+ + +	0
	3	− −	0	0
C	1	0	0	0
	2	− − −	0	0
	3	0	− − −	0

* The contrasts were set up as described in the text: effect A = 0 − saline; effect B = 36 − 0; effect C = 108 − 36. Significant positive differences are indicated by pluses; significant negative differences by minuses. One plus or minus represents $p < 0.10$, two symbols represent $p < 0.05$, and three symbols represent $p < 0.01$. Zero indicates no significant difference.

6.4. Discussion

As a pilot experiment to the study reported above, Stewart et al. (1970) gave donors 18 trials of training with either light or dark as the rewarded stimulus. Each brain was homogenized and then divided into doses equivalent either to 0.024 or to 0.145 parts brain. This allowed five recipients for each combination of dose and donor source. The recipients were tested in the blind over three consecutive sessions. The effect of donor training (or donor source) was significant only in the first session. The effect was negative transfer and did not depend on dose. The effect of dose was significant only in the last two sessions during which the higher dose group, regardless of donor training, performed better than the lower dose group.

In the present and larger scale experiment, dose and level of donor training were not simply additive and the interaction of these two factors changed over the course of recipient testing. This result provides some support for the concept that qualitatively different transfer agents may be extracted at different stages of donor training (Halstead and Rucker 1970; Rucker and Halstead 1970). In view of the non-monotonic dose–response relations reported by Rosenblatt (Rosenblatt and Miller 1966; Rosenblatt 1970a; chapter 3, this volume), it would be necessary to test a much larger number of combinations of dose and level of donor training to show conclusively that these two factors are not equivalent and do not represent the same thing on, perhaps, different scales. The technique of mixing aliquots of extract from donors receiving different training, used by Cartwright (1970), may offer another approach to the question of whether dose and level of donor training are completely exchangeable.

References

BROWNLEE, K. A., 1965, Statistical theory and methodology, 2nd ed. John Wiley and Sons, Inc., New York, pp. 225 and 316.

CARTWRIGHT, G. M., 1970, J. Biol. Psychol. *12*, 53.

ELLIS, H. C., 1965, The transfer of learning. The Macmillan Company, New York.

FJERDINGSTAD, E. J., T. NISSEN and H. H. RØIGAARD-PETERSEN, 1965, Scand. J. Psychol. *6*, 1.

GOLUB, A. M., F. R. MASIARZ, T. VILLARS and J. V. MCCONNELL, 1970, Science *168*, 392.

HALSTEAD, W. C. and W. B. RUCKER, 1970, The molecular biology of memory. *In:* W. Byrne, ed., Molecular approaches to learning and memory. Academic Press, New York, pp. 1–14.

MACKINTOSH, N. J., 1969, J. Comp. Physiol. Psychol. Monograph *67* (no. 2, part 2), 1.

NISSEN, T., H. H. RØIGAARD-PETERSEN and E. J. FJERDINGSTAD, 1965, Scand. J. Psychol. *6*, 265.

ROSENBLATT, F., 1970a, Induction of discriminatory behavior by means of brain extracts. *In:* W. Byrne, ed., Molecular approaches to learning and memory. Academic Press, New York, pp. 195–242.

ROSENBLATT, F., 1970b, Induction of behavior by mammalian brain extracts. *In:* G. Ungar, ed., Molecular mechanisms in memory and learning. Plenum Press, New York, pp. 103–147.

ROSENBLATT, F., 1970c, Effects of trained brain extracts on behavior. Paper presented for symposium on the Chemistry of Memory, chaired by G. Adam and sponsored by the Hungarian Academy of Sciences, Tihany, Hungary, September 1–5, 1969 (under publication by Akademiai Kiado, Budapest).

ROSENBLATT, F. and R. G. MILLER, 1966, Proc. Natl. Acad. Sci. U.S.A. *56*, 1423 and 1683.

RUCKER, W. B. and W. C. HALSTEAD, 1970, Memory: antagonistic transfer effects. *In:* W. Byrne, ed., Molecular approaches to learning and memory. Academic Press, New York, pp. 295–306.

SCHAD, L. J., J. E. ROLLINS and C. W. SNYDER, 1969, Psychon. Sci. *14* (3), 112 and 114.

STEWART, M., D. STRACHMAN and W. B. RUCKER, 1970, Paper under preparation for Matrix.

UNGAR, G. and L. N. IRWIN, 1967, Nature *214*, 453.

CHAPTER 7

A derepressor hypothesis of memory transfer

STANISLAV REINIŠ*

Department of Physiology, Ghana Medical School, Accra, Ghana

7.1. Introduction

The changes in the central nervous system leading to the formation of the permanent memory trace are at present unknown. We are not able to localize satisfactorily the phenomena accompanying the choice, transformation and fixation of information either in the macrostructure or in the microstructure of the brain.

One of the methods which may contribute to the solving of the problem of memory mechanisms is so-called memory transfer – facilitation of learning by the injection of extracts from trained brains. In the present paper, we have tried to explain the mechanism of the effect of substances responsible for memory transfer. We did not study the structure of these substances or the specificity of the effect.

7.2. Facilitation of alimentary conditioning by brain extracts from trained animals

Our first experiments were performed with white rats of the Wistar strain. Classical alimentary conditioning, presenting food pellets associated with light or sound, was used for the testing of the memory transfer. The food was placed in the food magazine, covered with a semitransparent door, the light and loudspeaker were located in the same wall of the conditioning box. The apparatus was equipped with a simple programming device, and the course of the experiment was therefore independent of the experimenters. Several

* Present address: Department of Psychology, York University, Toronto, Ontario, Canada.

aspects of the activity of the animals in the experimental box were followed. At the beginning of each experiment the time period necessary for orientation in the experimental box was measured. This period of exploratory activity ran from the time of introduction of the animal into the box to the first opening of the door covering the food. Before introducing the animal into the box, the food (a piece of Larsen diet) was already present in the food cup. In each experimental session the rats had ten trials. The duration of the conditioned stimulus (light from 15 W bulb placed next to the food cup or, in other group of experiments, sound of 2000 c/s) was maximally 20 sec. When the rat opened the door of the food cup during the conditioned stimulus, the conditioned stimulus was automatically terminated. The interval between trials was 2 min.

In addition to the orientation time in the experimental box and the number of positive responses, the latency of positive responses and the number of intertrial responses were measured.

The donor rats received 300 trials in 30 experimental sessions. Brains of rats that had 80% correct responses and fewer than one intertrial reaction per min during the last three experimental sessions were used for i.p. injection.

About 1 hr after the last experimental session control rats were decapitated, the brain was quickly removed, weighed and homogenized in a glass homogenizer in 3 ml of physiological saline. The brain hemispheres, the diencephalon and part of the mesencephalon were homogenized, i.e. that part of the brain divided by section at the level of tentorium cerebelli. Cerebellum, rhombencephalon and olfactory lobes were not used.

The suspension was injected i.p. into experimentally naive rats immediately after homogenization. The injection was performed through a wide needle (40 mm length, 1.2 mm width). Each brain was treated separately and each recipient animal received the homogenate of one trained brain. No external bleeding was observed after injection. The injected rats were divided into two groups. The first group was starved from injection to first experimental session, i.e. for 48 hrs. The other group was deprived for only 16 hrs before the first experiment.

The animals injected with brain homogenate of trained rats were compared with two control groups. The first control group was injected with the brain homogenate of experimentally naive rats. The other control group received homogenate of naive rat liver (1500 mg in 3 ml of physiological saline). These control groups were also divided into two sub-groups, one starved for

16 hrs before the first session, the other for 48 hrs. All groups injected with homogenate were given 15 experimental sessions.

The results of this group of experiments showed that the homogenates of the brains of trained animals affect substantially the performance of injected animals in the same experimental situation (see table 7.1).

After being placed in the experimental box, rats injected with the homogenate of the trained brain found the food cup much quicker than control animals injected with the homogenate of liver or untrained brain, i.e. the exploratory period was much shorter.

The number of positive conditioned responses (opening of the door covering the food cup during the conditioned stimulus) was also substantially higher in rats injected with homogenates of trained brains than in any of our control groups. The number of intertrial reactions was not higher than the number of intertrial reactions in rats injected with the homogenate of untrained brains. This showed that our results were not caused by a non-specific increase of general motor activity only, but had some relation to the conditioning.

However, the animals which were fed between the i.p. injection of the brain homogenates and first testing session did not differ from the control animals (see table 7.2). Also the performance in the experimental box was not improved in these animals.

We did not explore the reasons why the memory transfer is present only in deprived animals. Perhaps this phenomenon is connected with the absorption of the material from the peritoneal cavity, with some metabolic changes during starvation; or with changes of the level of some ions, calcium, potassium or some other factor. Maybe positive motivation is somehow necessary for the effect of homogenates of trained brain.

7.3. Labelling of brain homogenates with radioactive phosphorus

The described phenomenon was called by other authors 'memory transfer' and was related very often to the presence of special types of ribonucleic acids which were supposed to be produced in the brain during learning. This RNA was said to be extracted from the brain of trained animals and transferred to the brains of recipients. 'Memory RNA' therefore should infect the brains of recipient animals and 'tune' them to the performance pattern

TABLE 7.1

Orientation time, number of conditioned responses and intertrial reactions in the first experimental session in naive rats injected with brain and liver homogenates (starving group).

Type of injection	Time to find food	Number of conditioned responses with conditioned stimulus		Number of intertrial reactions per min with conditioned stimulus	
		Light	Sound	Light	Sound
No injection	699 ± 99 sec	14.7 ± 5%	20.5 ± 6%	0.78 ± 0.28	0.83 ± 0.35
Naive brain	526 ± 82 sec	22.0 ± 11%	23.6 ± 8%	2.27 ± 1.18	2.49 ± 0.60
Trained brain	114 ± 16 sec	63.3 ± 8%	55.7 ± 8%	2.12 ± 0.37	2.80 ± 0.51
Liver	527 ± 127 sec	27.8 ± 11%	18.3 ± 9%	1.54 ± 0.69	0.79 ± 0.23

TABLE 7.2

Orientation time, number of conditioned responses and intertrial reactions in the first experimental session in naive rats injected with brain and liver homogenates (non-starving group).

Type of injection	Time to find food	Number of conditioned responses with conditioned stimulus		Number of intertrial reactions per min with conditioned stimulus	
		Light	Sound	Light	Sound
No injection	1076 ± 218 sec	19.0 ± 8%	24.0 ± 9%	1.72 ± 0.76	2.10 ± 0.95
Naive brain	617 ± 242 sec	11.4 ± 7%	8.3 ± 4%	0.94 ± 0.43	0.20 ± 0.18
Trained brain	765 ± 198 sec	17.1 ± 4%	19.2 ± 5%	0.56 ± 0.32	0.17 ± 0.09
Liver	1125 ± 164 sec	21.3 ± 9%	17.1 ± 9%	1.55 ± 1.13	1.57 ± 0.68

which was previously produced in the brains of trained animals. The molecule of such memory RNA should code the obtained information.

The first question which we asked was: does RNA really penetrate into the brain tissue of the recipient animals? And if it can pass the blood–brain barrier, is the amount of this RNA sufficiently high to be detected?

We tried therefore to label with radioactive phosphorus the RNA formed during training and then, after the injection of such brain homogenate to the recipient animals, to detect the traces of the radioactivity in the brains of recipients.

In the period of maximal increase of the number of correct conditioned responses, i.e. during the first training sessions of the donors, radioactive phosphorus in the form of disodium hydrogen phosphate in physiological saline was i.p. given to the donor animals. The first dose was injected 3 hrs before the first session, the second dose given before the fourth session and the third dose before the seventh session. Each animal received a total of 150 μC of radioactive phosphorus per 100 g of weight.

After 30 experimental sessions, the animals were killed, again by decapitation, and the brains homogenized.

One-half ml of the homogenate was dried and its radioactivity measured. The rest (3.5 ml) of the suspension was injected i.p. into the recipient animals. These recipient rats were starved again for 48 hrs after injection, and then tested in the experimental box. After another ten sessions, the recipient rats were killed, their brains were removed, weighed and homogenized, and the radioactivity was measured in 4 ml of the homogenate. In this group of ten animals, we also found memory transfer (see table 7.3).

TABLE 7.3

Reaction of control rats and rats injected with the homogenate of trained or untrained brains in the first (testing) experimental session.

Animals	Time to find food in sec	Conditioned reactions in %	Intertrial reactions per session
Controls	812 ± 106	20.5 ± 6	15.6
Injection of untrained brain	726 ± 82	23.0 ± 8	45.4
Injection of trained brain	102 ± 18	58.0 ± 8.0	42.4

In the brain homogenates of donor animals (table 7.4), we detected a relatively high radioactivity. We did not analyze the distribution of radioactive phosphorus into individual fractions, and we therefore did not know how much RNA was actually labelled, nor how much ^{32}P was incorporated into phospholipids or any other substance.

TABLE 7.4

Radioactivity of the rat brain and liver homogenates of donor and recipient animals (values are given in number of counts per min minus background).

	Brain	Liver
Donors – on the day of injection	614 ± 23	245 ± 30
Recipients – 12 days after the injection of homogenate	No detectable activity	No detectable activity

After the injection of this labelled homogenate, we did not find any radioactivity in the brains of recipient animals which exceeded the background activity. This experiment has only orientative value, since it was not known whether there was any labelled RNA in the brain homogenate injected to the recipient animals at all. We did not analyze these facts further, and we did not relate obtained values to total RNA, protein, or phosphorus of the sample, because in any case the result of the experiment is negative. Wherever radioactive phosphorus is incorporated, one fact is sure, that no radioactivity was found in the brains of the recipients.

Another possible explanation of the failure of this experiment is that the labelled RNA, bearer of the obtained information, really penetrated into the brains of recipient animals, induced biochemical changes leading to the formation of a stable memory trace, and was then quickly eliminated. There are other possible objections to the value of this experiment. Thus, the only possible conclusion to be drawn from the experiment is that no substance labelled by radioactive phosphorus during learning was permanently incorporated into the brains of recipient animals. The memory transfer therefore does not involve a stable incorporation of detectable amounts of any substance containing ^{32}P which is labelled at the time of training, and therefore memory transfer probably does not involve the stable incorporation of 'memory RNA'.

7.4. Transfer of old, fixed memory traces

Memory transfer or 'transfer of the memory trace' is, as the term itself implies, sometimes supposed to be direct 'transplantation' of information stored in the brain of the donor into the brain of recipient animal. However, the brains of trained animals do not contain only the information stored during our experimental procedure. Even the brains of untrained animals which we used as controls are not totally devoid of information. Since the moment of birth these animals have learned how to live in their environment, to know other animals in the cage etc. These brains store also the inborn 'information' which determines the inborn behavioral pattern of the animal.

Does memory transfer mean that we really transfer all this? Are we really able to transfer the whole genotype and phenotype of the donor to the recipient, any information inborn or obtained during the life of the individual?

If this hypothesis were correct, then any memory trace, recent or very old, should be transferable.

We analyzed this hypothesis by the following experiment: some rats are found which consistently and 'spontaneously', i.e. without previous training, kill mice, whereas other rats of similar genetic and environmental background never kill them. The reason for the difference between these two groups of rats is not understood. However, killers do not spontaneously revert to non-killers (Myer 1966). This behavior is supposed to be innate (Karli 1961) or dependent on early experience (Myer 1969). In any case, the behavior is not recently learned. There might be some similarity between the biochemical mechanism of the storage of inborn and of learned behavior. Both of them may be somehow fixed in the system of nucleic acids, and both of them may be transcribed from there into proteins which alter the function of groups of nerve cells. We are far from being able to recognize what is similar in the storage of these two types of behavioral pattern or where the differences are. For the purpose of this experiment, let us suppose that there is really some similarity in the method of storage.

In a sample of 48 rats, 14 rats were found that consistently killed mice. The rats were kept for three days in individual cages and after three days were tested for mouse killing behavior by being given individually a single mouse for 1 hr on three successive days. This method of identification of killer rats has previously been used by another author (Myer 1966). After three days of testing eight of the killer rats were quickly decapitated and their brains homogenized in exactly the same way as in previous experiments. The brain homogenate from individual killer rats was injected i.p. into a non-killer rat. The eight injected non-killers were starved for 48 hrs after injection. The procedure was therefore identical with that previously proved successful in the positive transfer of training. After starving, the injected recipient animals were tested for the mouse-killing response and were then retested again at weekly intervals for a period of four weeks. We also tried the same experiment in the opposite way. We injected the brain homogenates of non-killers to killer rats. Four non-killer rats were decapitated, their brains homogenized and injected individually into four killer rats. After injection, the recipient animals were starved for 48 hrs and were then tested for mouse killing.

None of the rats changed their behavior. Killers remained killers, non-killers remained non-killers.

We deduced from this experiment that a behavior which is probably either inborn or learned a long time ago cannot be transferred. There is uncertainty to whether mouse killing by rats is an inborn or an acquired type of behavior and whether there is some similarity in the storage of learned and inborn behavior at all. Therefore we performed a similar experiment with the transfer of an older memory trace.

For this purpose, we modified our experiments with transfer of the alimentary conditioned reflex, using albino mice. Albino mice were trained in a conditioning box with 29×18 cm floor and 12 cm high walls. In the shorter wall of the chamber, a swinging door 3×5 cm was placed which covered a small food magazine. An amber light above the transparent swinging door acted as a conditioned stimulus. The conditioned stimulus lasted for 10 sec during which a food pellet was dropped into the food magazine. If the animals took the food during the conditioned stimulus, the conditioned response was considered positive. The period between conditioned stimuli was 80 sec, with ten trials in one session. Mice were considered to be trained when they gave ten positive responses and showed fewer than 20 intertrial responses in one session. Opening of the swinging door in the period between conditioned stimuli was considered as intertrial response.

When we tested the effect of brain homogenates from trained mice on the behavior of naive mice, we found that whole homogenates had a toxic effect and usually killed the recipients. It was therefore necessary to prepare brain extracts. The mice were decapitated and the brains were quickly removed and homogenized in a glass homogenizer for 2 min with 2 ml of distilled water. The homogenate was shaken for 4 hrs, left in the refrigerator at 1–3 °C overnight and centrifuged at 4000 rpm for 2 hrs. All procedures with the exception of the extraction in the refrigerator were performed in a room with a temperature of 22–24 °C. The extracts were injected i.p. into experimentally naive animals which were not fed on the day of the injection or on the following day of testing. The recipient animals were tested 24 hrs after the injection.

We did not prepare the extracts in the cold as other authors do. When we used the whole homogenates, we injected them i.p. The homogenate in the peritoneal cavity was therefore for several hours under conditions similar to the conditions in a 38 °C warm incubator, the compounds responsible for the memory transfer being affected by the enzymes of the brain homogenate as well as by the enzymes of the body of the recipient. The active substances must have passed through several barriers, they were transported by the blood stream – and in spite of that they remained active. They were not therefore very labile.

Thus we did not suppose that the active substances would be damaged too much by the extraction at room temperature. Why handle the trained brains on dry ice etc., when the extracts are then exposed to all hydrolytic and other enzymes of the body of the recipient?

Our extracts prepared at room temperature were really active. We used these extracts in several groups of experiments and in different conditions. In table 7.5 is presented the survey of mean numbers of positive conditioned

TABLE 7.5

Summary of all memory transfer experiments performed in 1967/68 in our laboratory, using alimentary conditioning of mice. UCS = food pellet, CS = light. Data are from first (testing) session.

Recipient group	Mean number of positive conditioned responses	Mean number of intertrial reactions per conditioned response
Controls	1.28	16.6
Injection of naive brain extract	1.69	15.4
Injection of trained brain extract	5.45	3.4

responses in the first experimental session after the injection. We did not follow the effects of liver homogenates any longer; we compared only control animals without any injection with animals injected with extracts of trained and untrained brains. In table 7.6 the mice are classified according to the number of positive conditioned responses in the first session. The difference between

TABLE 7.6

Distribution of numbers of conditioned responses in the first (testing) session after injection of brain extracts in mice (N – animals which did not find the food cup hidden behind the transparent door at all).

No. of responses	Controls		Injection of naive brain		Injection of trained brain	
	abs.	%	abs.	%	abs.	%
N	78	26.0	9	19.8	2	4.08
0	98	32.6	5	11.0	2	4.08
1	41	13.7	8	17.6	0	–
2	45	15.0	9	19.8	2	4.08
3	18	6.0	6	13.2	1	2.04
4	7	2.3	9	19.8	6	12.24
5	8	2.7	0	–	10	20.40
6	5	1.7	0	–	11	22.44
7	0	–	0	–	8	16.32
8	0	–	0	–	5	10.20
9	0	–	0	–	1	2.04
10	0	–	0	–	1	2.04
No. of animals	300		46		49	

the animals injected with extracts of trained brains and the control groups is highly statistically significant. All values are summarized from the transfer experiments described in further parts of the present paper, with the exception of experiments combining memory transfer with application of drugs.

In the first group of experiments we tested the transfer of older fixed memory trace. The donors were divided into four groups: the first group consisted of experimentally naive, untrained mice. The second group consisted of trained mice who gave ten conditioned responses in one session, with the number of intertrial responses being very high, between 60 and 80 in one session. The number of training sessions in this group was 12–14. The third group of donors, killed after 22–26 sessions, had ten conditioned responses in one session and fewer than ten intertrial responses. A fourth group of donors was fully trained to the criterion (in 22–26 sessions), then left 15 days without any training. After 15 days they were once again tested in the conditioning chamber and killed. The brain extracts were then injected again into the naive recipient mice. The donors, after 15 days of rest, showed almost complete retention, average number of conditioned responses being 9.53.

The results are summarized in table 7.7. Injections of extracts from untrained brains as well as from incompletely trained brains produced no subsequent change in the behavior of the recipient animals in the experimental

TABLE 7.7

Number of conditioned responses in the testing session in mice after injection of brain extracts from mice trained to different degrees.

Donors	Average number of positive responses
Untrained	2.70
Incompletely trained	2.50
Completely trained	6.25
Killed two weeks after completed training	2.12

chamber, while injections of brain extracts from the completely trained mice of the third group produced a significant increase in the number of conditioned responses in the first experimental session of untrained recipients. The brain extracts of donors killed two weeks after completed training were fully ineffective, the performance of recipients in the first session being 2.12 conditioned responses.

This experiment therefore shows again that memory transfer is a phenomenon connected with the process of training, not with the already fixed memory trace. What we transfer is not the substance in which the ready information is permanently coded. We transfer something which is produced in the brain during learning, which probably has some physiological importance for learning, and which disappears when the learned information is already fixed. The term memory transfer is therefore misleading because we are not able to transfer an older stabilized memory trace. Only brains with a fresh, recent memory trace are active and able to modify the behavior of recipient animals. This fact has been shown also in the papers of other authors (Rosenblatt 1966; McConnell 1967; Ádám and Faiszt 1968).

7.5. Combination of memory transfer with puromycin

What is the probable importance of substances responsible for memory transfer?

For a possible explanation, we used the model of memory storage proposed by E. Roy John (1967). John's model of the memory storage in the individual nerve cell supposes that increased synaptic transmission connected with increased number of afferent impulses during learning activates repressed, inactive parts of the genome. Induced production of RNA causes increased protein synthesis. New proteins probably somehow alter the synaptic transmission by inducing the production of an increased amount of neurotransmitter, or by changing the functional properties of the postsynaptic membrane. The alteration of the activity induces further synthesis of derepressors, which activate a part of genome (which might therefore be called 'memory operon'). Thus, RNA production is further increased which is a template for the production of new protein. This protein increases further the synaptic activity, and more derepressors are produced up to the moment when the derepression of the memory operon is completed. Then the production of derepressors may gradually decrease.

This is only a hypothesis almost all of whose points have to be still proved. However, we have used this hypothesis for the determination of the role of the substances of memory transfer. We did not suppose that newly formed RNA was the transfer substance. The above described experiments with radioactive phosphorus, the findings of several authors (Rosenblatt et al. 1966; Ungar and Irwin 1967; Chapouthier 1968) that the active substances are possibly peptides, the impossibility of the passage of unchanged RNA through the blood-barrier (Luttges et al. 1966; Corson and Enesco 1966) show that RNA is probably not directly involved in memory transfer.

Numerous experiments with memory transfer using RNA-rich extracts have been presented (Jacobson et al. 1965; Fjerdingstad et al. 1965; Gay and Raphelson 1967); however, it is not certain that these extracts did not contain other substances with transfer activity. On the other hand, it is also possible that RNA acts in the body of recipients by some other, indirect, way.

At this stage of experiments, we were not able yet to exclude the possibility that RNA is the active substance. However, in our extracts prepared at room temperature there would hardly be any undegraded RNA which could be a template for any complicated protein molecule. Therefore, we did not suppose that RNA is the main factor responsible for memory transfer.

There are two other possibilities which we investigated. Substances responsible for memory transfer might be either newly produced specific proteins which may be extracted from the tissue, and which may play similar roles (probably activation of the synaptic transmission) in analogous areas of the brain of the recipient; or they may be derepressors which, after the injection into the body of the recipient, activate the DNA of some nerve cells of the brain.

These two possibilities were examined by the combination of the memory transfer with antibiotics affecting protein metabolism.

The antibiotics were injected intracranially either into donors or into recipients. The mice were slightly narcotized with ether and the skin of their heads cut in the midline. A thin intradermal needle was inserted into the skull through the temporal muscle. The solution of the substance was then injected to each hemisphere by means of a Hamilton microsyringe. We tried to avoid damage to the brain tissue, administering the drug into the subdural space. The spreading of the solutions over the surface of each hemisphere was easily observed through the semitransparent skull. The whole visible cortex was in contact with the solution.

The first tested drug was puromycin. Puromycin is an antibiotic that blocks the protein synthesis in ribosomes. Puromycin is bound to the peptides and causes their premature release from the ribosomes. Using this antibiotic we tried to decide whether the 'memory transfer substance' is a protein already formed in donor ribosomes during learning and extracted from the brains of the donors before the fixation somewhere in the synapse. If it were this protein, then puromycin injected to the recipients should be ineffective. Puromycin was injected intracranially in the dose 40 μg in 60 μl of distilled water.

The experiments are summarized in table 7.8. Puromycin injected intracranially into the recipients simultaneously with the extract of trained brain inhibits the transfer effect. In this group of experiments the only positive

TABLE 7.8

Conditioned responses in the first experimental session in recipients injected with brain extracts i.p. plus saline or puromycin intracranially.

Donors	Intracerebral injection to the recipients	No. of animals	Average no. of conditioned response in the 1st session	Average no. of intertrial reactions/session	Av. no. of intertrial reactions/one conditioned response
No donors	Saline	11	1.82	28.1	15.4
No donors	Puromycin	18	0.17	9.5	55.8
Untrained	Saline	10	1.40	21.1	15.1
Untrained	Puromycin	13	1.69	20.1	11.8
Trained	Saline	10	4.35	15.1	3.4
Trained	Puromycin	12	1.17	23.8	20.3
No donors	None	300	1.28	16.6	12.9

transfer effect was found in recipient mice injected with saline intracranially and with extract of trained brains i.p. This group differs significantly from all other groups summarized in table 7.8, i.e. naive mice injected intracranially with saline or puromycin without any i.p. injection, naive recipients injected with saline or puromycin intracranially and with extract of the untrained brain i.p., and naive recipients injected with puromycin intracranially and extract of the trained brain i.p.

This experiment may therefore show that protein newly formed during learning, extracted from the brain of the donor, and supposedly fixed to the corresponding synapses of the recipient is *not* the active substance. If this were so, memory transfer should be present even in the recipients affected simultaneously by puromycin and extracts of the trained brains. The second possibility, memory transfer substances acting as derepressors of DNA in the brains of recipient individuals, seems after this experiment more probable. This experiment, on the other hand, does not exclude the possibility of RNA as the active substance.

Puromycin is not a very suitable drug for this type of experiments. It induces the production of incomplete, defective peptides which, under certain conditions, may themselves affect the performance of trained animals (Flexner et al. 1964). We showed this also in the following experiment: we trained the mice in the same experimental situation as described above. In 20 sessions we relatively firmly fixed the alimentary conditioned reflex. The second group of mice was not trained at all. These mice were slightly narcotized by ether and 20 μg of puromycin dissolved in 30 μl of distilled water was injected intracranially above each hemisphere. This dose was sufficient for the blockage of performance of trained animals. It was found in other experiments, that when the trained animals received 40 μg of puromycin intracranially, their performance decreased from ten conditioned reactions in one session to 0.78. I.p. injection of the same dose was ineffective.

Twenty-four hrs after the injection of puromycin into the untrained animals, the mice were sacrificed by decapitation and extracts of these untrained puromycin brains were prepared in the usual way.

Clear supernatant of the centrifuged homogenate of the brains of puromycin-affected animals was injected i.p. into the trained mice. The extract from one brain was always injected into one recipient animal. After the injection, the trained mice were deprived of food for 24 hrs and then were tested for the performance in the experimental chamber. The following control groups were tested and compared with the main studied group:

(1) Trained mice injected with the extracts of saline-affected brains. Instead of puromycin, 30 μl of physiological saline was injected to each hemisphere of the untrained animal; otherwise, the procedure was identical with the one described above.

(2) Untrained mice injected with the extracts of saline-affected untrained brains.

(3) Untrained mice injected with the extracts of puromycin-affected untrained brains.

(4) Untrained mice without any injection placed for the first time into the experimental chamber.

TABLE 7.9

Number of conditioned responses after the application of extracts of brains treated by puromycin.

Intracerebral injection to the donor	Recipient	No. of recipient animals	Average no. of conditioned responses before injection of brain extract	Average no. of conditioned responses after injection of brain extract
Puromycin	Trained	12	10.00	1.83
Saline	Trained	10	10.00	10.00
Puromycin	Untrained	16	–	1.19
Saline	Untrained	12	–	1.58
No donor	Untrained	22	–	1.27

The results are shown in table 7.9. I.p. injection of the brain extracts from saline-treated untrained brains does not affect performance of the trained animals at all. By contrast, an extract from puromycin-treated untrained brains reduces the performance of trained animals almost to the level of untrained ones. There is a significant difference between trained animals injected with saline-treated brain extracts and puromycin-treated brain extracts (Wilcoxon test, $S=120$, $p<0.0001$). There are no significant differences between untrained animals without any injection, and animals injected with extracts from puromycin-treated and saline-treated brain.

This experiment may be evidence of the activity of defective peptides produced in the brain under the effect of puromycin. These defective peptides may also block the performance of the animals affected by extracts of trained brains combined with puromycin.

The effect of puromycin is probably rather complicated. In further group of experiments, we injected puromycin into the brains of trained donors and investigated the effect of puromycin-treated brain extracts on the performance of naive animals.

We used again the same method of training. We had the following groups of donors:

(1) Donors trained, intracranial injection of puromycin. Recipients untrained.

(2) Donors trained, intracranial injection of saline. Recipients untrained.

(3) Donors untrained, intracranial injection of puromycin. Recipients untrained.

(4) Donors untrained, intracranial injection of saline. Recipients untrained.

Table 7.10 shows the results of this experiment. Intracranial injection of 40 μg of puromycin inhibits the performance of trained animals in the experimental chamber. The number of conditioned responses decreased from 10.00 to 0.78. The injection procedure, involving narcosis etc., may also be partly responsible for this decrease. If only saline was injected into the trained animals intracranially, their performance also decreased from 10.00 to 7.88 within 24 hrs.

The brain extracts from animals treated with puromycin were i.p. injected into the untrained mice. Testing session after 24 hrs showed that memory transfer is present also in this case.

There is no statistically significant difference between animals injected with extracts of puromycin- and saline-treated trained brain. There are only significant differences between animals injected with extracts of puromycin-treated trained and untrained brains as well as between animals injected with extracts of saline treated trained and untrained brains.

Puromycin injected to the donor therefore does not affect the memory transfer. It might also be an evidence that protein newly formed during learning is not the substance performing memory transfer. Under the effect of puromycin defective peptides are formed instead of this protein. In spite of this production of defective peptides, memory transfer is still possible, because the derepressors are probably not immediately and directly affected by puromycin.

Similar results were obtained also with methionine-sulphoximine. In our

TABLE 7.10

Conditioned responses in the first experimental session in untrained recipients injected with extracts of puromycin- or saline-treated brains.

Donors	Average no. of conditioned responses before the injection	Average no. of conditioned responses after the injection	No. of recipients	Average no. of conditioned responses in the 1st exp. session	p	Average no. of intertrial reactions per session	Average no. of intertrial reactions per one conditioned response
Untrained, intra-cranial injection of saline	–	–	12	1.58		24.5	15.5
Trained, intra-cranial injection of saline	10.00	7.88	8	4.63	< 0.005	15.9	3.4
Untrained, intra-cranial injection of puromycin	–	–	16	1.19		16.9	14.2
Trained, intra-cranial injection of puromycin	10.00	0.78	9	5.55	< 0.001	10.9	1.9
Untrained animals without any injection			119	1.17		16.6	12.9

experiments which are not described here in detail, because they do not fit exactly into the schedule of this paper (Reiniš and Kolousek 1968), we found that methionine-sulphoximine blocks the performance of trained animals, but brain extracts from these animals still also have memory transfer properties.

The relation of puromycin, memory transfer and the postulated presence of defective proteins should be further elucidated by other modifications of similar experiments.

7.6. Block of memory transfer by actinomycin D

Another piece of evidence for the derepressor hypothesis was obtained by the combination of memory transfer with the effect of actinomycin D. Actinomycin D is an antibiotic blocking the activity of RNA polymerase and therefore the production of new RNA.

We injected actinomycin D in the dose of 1 μg dissolved in 20 μl of physiological saline intracranially into the recipient animals. The drug was injected immediately before the i.p. administration of the extracts of the trained or untrained brains.

The donor mice were again trained in the same experimental situation. An alimentary conditioned reflex to light was used. In addition to the principal experimental group, which was injected intracranially with 1 μg of actinomycin D and i.p. with extract of trained brain, the following control groups of animals were also tested:

(1) Control mice without injection.
(2) Mice injected with 1 μg of actinomycin D only.
(3) Mice injected with 0.02 ml saline to each hemisphere only.
(4) Mice injected intracranially with 1 μg of actinomycin D plus extract of untrained brain i.p.
(5) Mice injected with 0.02 ml saline intracranially plus extract of untrained brain i.p.

The dose of 1 μg of actinomycin D did not alter the motor activity of the animals on the rotating rod, nor the explorative activity of animals in the girdle cage, and it only slightly decreased the food consumption at least 24 hrs after the intracranial injection. We tested the effect of this dose on mentioned functions in control experiments.

The results summarized in table 7.11 show that groups of control animals without any injection, animals injected with physiological saline or 1 μg of actinomycin D intracranially only, and animals injected with a combination of saline or actinomycin D intracranially with extracts of untrained brain i.p., do not differ substantially from each other. The slight differences in performance during the testing sessions are not statistically significant.

TABLE 7.11

Response of recipient mice to the testing session in the experimental chamber after the injection of brain extracts and actinomycin D.

Group	Intracranial injection	Intraperitoneal injection of the extract	No. of animals	Mean no. of conditioned responses in the session
1	–	–	119	1.17
2	Saline	–	11	1.82
3	Actinomycin D	–	16	1.25
4	Saline	Untrained brain	10	1.40
5	Actinomycin D	Untrained brain	8	1.13
6	Saline	Trained brain	10	4.10
7	Actinomycin D	Trained brain	9	1.56

The mice injected intracranially with physiological saline and i.p. with extract of brains of trained animals show significantly higher performance than other groups. This represents the positive transfer effect in this group of experiments. The positive transfer effect is, however, blocked by the simultaneous intracranial injection of actinomycin D. The difference is statistically significant ($S = 52$, $p < 0.02$) calculated by the Wilcoxon test.

Our experiments show that the behavioral modification obtained with extracts from the brains of recently trained animals may be caused by the effect of active substances present in the extracts on the formation of new RNA on DNA templates. If the formation of new RNA is blocked by actinomycin D, brain extracts are ineffective. Thus these experiments support the theory of the derepressor character of substances responsible for memory transfer.

We may assume that the derepressors are involved probably also in the normal formation of the memory trace. It may be supposed that they may diffuse from the place of origin not only to the nucleus of the nerve cell engaged in learning, but also to the neighboring glial cells and to other nerve cells. The derepressors may in this case be polypeptides and/or small proteins easily penetrating through the membranes.

7.7. Pilot experiments with hydroxylamine, a mutagen affecting activated DNA

The derepressor hypothesis of memory transfer implies that even during normal learning, without the effect of brain extracts or homogenates, the derepressors may diffuse through the brain tissue and activate larger areas of the brain. The memory transfer may only simulate this phenomenon.

One important question should be solved for the verification of this hypothesis: is DNA of the nerve cells really derepressed during learning? Does it remain derepressed after the learning is completed? We tried to contribute to the solution of this problem by a series of experiments in which we did not use the method of memory transfer. These experiments are, however, closely related to the problems connected with memory transfer and should therefore be mentioned here.

Hydroxylamine, NH_2OH, is known not only as a potent inhibitor of protein synthesis (Rosenkranz and Bendich 1964), and the synthesis of RNA, but also as a mutagen. In vitro, hydroxylamine is highly specific for the transition of cytosine into N_6-hydroxy-4,5-dihydro-4-hydrosylaminocytosine, which is later transcribed as if it were thymine (Freese et al. 1961). This specificity is lost in vivo, presumably because of the multilateral effects of hydroxylamine on cellular metabolism (Tessman et al. 1965). Some other changes of nucleotides have also been described. This drug is able to cause mutations not only in bacteriophage (Tessman 1966) and *E. coli* (Weigert and Garen 1965), but also in mammalian cells (Borenfreund et al. 1964). Under the effect of hydroxylamine, proteins with changed sequence of amino acids are formed, which therefore lack the functional specificity of normal proteins.

It is important for our experiments that hydroxylamine affects more easily activated, derepressed DNA (Freese and Freese 1964) leaving alone the inactive, repressed one.

In the first experiment with hydroxylamine, we used mice trained in a simple Y-maze filled with water at 24 °C. The starting arm was 30 in long and 5 in wide, and the two target arms were 18 in long and 5 in wide. In one arm, there was an escape platform 5 × 3 in with stairs. The angle between the target arms was 120°. The whole equipment was made from copper sheets, and the interior surfaces were painted black. In the pretesting period, preference of the mice for the right or left arm was determined. Escape platforms were placed in both arms, and the mice were run ten times in one session. The mice with left prevalence were then trained to escape to the right arm and vice versa. If in pretesting a mouse chose each arm five times we trained such an animal to the left or right side on a random basis. The animals which chose one arm ten times were eliminated from the experiment, because they reached criterion on the learning task significantly later than all other animals.

The interval between trials was 15 sec. If the mouse did not find the escape platform in 120 sec it was removed from the maze for 15 sec. Then the experiment continued.

The mice were trained to a criterion of one error out of five trials. The animals which reached criterion in fewer than eight trials were eliminated from final evaluation, because in the second (testing) session, they usually did not show retention. Mean number of trials to criterion was 14.0. Pairs of animals were chosen which reached the criterion in the same number of trials. One of them was injected with hydroxylamine and the other one with saline. Animals that died after the injection were replaced by animals with the same performance.

This matching according to performance of individual animals in groups was not possible in the animals injected before the training session. These groups were therefore chosen randomly.

Retention of the learned task was tested in one group of animals 24 hrs after the training session and in the second group after 48 hrs. Each animal was tested only once. The retention was expressed in a percentage figure calculated from the formula

$$\frac{(\text{Tr-5}) - (\text{Ts-5})}{(\text{Tr-5})} \times 100,$$

where Tr was the number of trials necessary for reaching criterion in the training session and Ts was the number of trials necessary for reaching criterion in the testing session. The results were evaluated by Wilcoxon's rank correlation analysis for dichotomic ranks. Hydroxylamine ($NH_2OH \cdot HCl$), neutralized by sodium hydroxide to pH 7.3, was injected intracranially in the dose of 20 μl. The concentration of hydroxylamine was 0.5 M.

The mice were slightly narcotized with ether and the skin of the head cut in the midline. An intradermal needle was inserted through the temporal muscle into skull. The injection through the temporal muscle as well as temporary compression of the muscle after the injection prevented leakage of the fluid from the skull. We tried again to avoid damage to the cerebral tissue and therefore injected the fluid into the subdural space, using a Hamilton microsyringe for the injection. The spreading of the fluid in this space was easily followed through the semitransparent skull. Ten μl of the fluid was injected toward each hemisphere in the temporal region. Since hydroxylamine-hydrochloride was neutralized by NaOH, 0.5 M saline represented roughly the solvent in which the hydroxylamine was administered. Therefore, 0.5 M saline was used for control experiments.

In this experiment, the effect of hydroxylamine injected at the following times was tested:

(a) 72 hrs before the training session
(b) 24 hrs before the training session
(c) 1 hr after the training session
(d) 4 hrs after the training session
(e) 24 hrs after the training session

The (a), (b), (e) groups of mice were tested 48 hrs after the training session. The experiments (c) and (d) were performed twice. One group of animals was tested 24 hrs after the training session, the second group was tested 48 hrs after the training session.

The results are summarized in tables 7.12 and 7.13. The animals injected with hydroxylamine 1 and 4 hrs after the injection and tested 24 hrs afterwards showed some impairment of retention. The differences between saline and hydroxylamine are statistically significant. The deterioration of performance is much more fully expressed in animals tested 48 hrs after the training session.

The mice injected 24 hrs after the training session and tested after another 24 hrs also showed some impairment of retention which was not statistically significant.

Impaired retention was observed also in animals injected 24 hrs before the training session ($p < 0.02$), but not in animals injected 72 hrs before the training session (see table 7.13).

TABLE 7.12

Percentage of retention of learning in the water maze after subdural injection of hydroxylamine following the training session.

Injection	Time of injection after the 1st session	Testing session after 24 hrs		Testing session after 48 hrs	
		%	No. of pairs	%	No. of pairs
Saline	1 hr	82.00	10	90.1	10
Hydroxy-lamine	1 hr	45.5		1.4	
Saline	4 hrs	79.9	10	97.4	8
Hydroxy-lamine	4 hrs	29.9		18.4	
Saline	24 hrs	–		78.0	9
Hydroxy-lamine	24 hrs	–		59.3	

TABLE 7.13

Percentage of retention of learning in the water maze after subdural injection of hydroxylamine 24 and 72 hrs before the training session (10 animals in each group).

	24 hrs	72 hrs
Saline	82.0	85.4
Hydroxylamine	33.5	71.6

The results indicate that hydroxylamine injected 24 hrs before the training session, and 1 or 4 hrs after the training session, impairs the performance of the animals in the testing session. The impairment is better expressed if the animals are tested 48 hrs after the training than if the animals are tested 24 hrs after the training only. Hydroxylamine injected 24 hrs after training has a smaller effect than if injected within the 4 hrs after learning.

At the end of the training procedure in the swimming maze some animals seemed to be exhausted. We did not find it useful to evaluate the performance of exhausted animals, i.e. if they had had more than 20 trials in one session. Therefore, one of the additional questions which we wanted to solve was whether or not the fatigue of the animals contributed to the impaired retention of the memory trace after hydroxylamine.

Therefore, for ten days before the training session in the water maze, animals swam in a big aquarium, on the first day for 5 min, and on nine successive days for 10 min. After ten days of physical exercise the animals were trained in the water maze, injected with 0.5 M saline or hydroxylamine exactly 1 hr after the training, and tested 48 hrs afterwards.

Pretrained animals injected with saline had in this experimental situation 77% retention, those injected with hydroxylamine had 13.6% average retention. The difference is highly significant statistically. These animals were seemingly not exhausted either at the end of training session or at the end of the testing session. Hence we assumed that exhaustion does not contribute substantially to the effect of hydroxylamine.

Another objection against the experiments with hydroxylamine may be that the effect of hydroxylamine is due to occult seizures. Cohen and Barondes (1967) found that occult seizures are probably responsible for the effect of puromycin on memory. This objection is not too plausible since hydroxylamine has, at least in some animal species, an anticonvulsive effect. However, an anticonvulsive effect of hydroxylamine has not yet been described in mice and therefore, for the elimination of this possibility, we used the method similar to that of Cohen and Barondes – we tried to block possible seizures by diphenylhydantoin.

The training procedure was again identical with the procedure described above. One hr after the training session, the mice were injected by hydroxylamine or saline. The testing session was carried out 48 hrs after the training. Four hrs before the testing session, the mice were injected with diphenylhydantoin in the dose of 35 mg/kg of body weight i.p.

Diphenylhydantoin did not alter substantially the effect of hydroxylamine. The mice injected with saline and diphenylhydantoin showed 94.6% retention, the mice injected with hydroxylamine and diphenylhydantoin showed 12.6% retention. This experiment therefore did not confirm that the effect of hydroxylamine on retention of memory trace is due to occult seizures.

Hydroxylamine has numerous effects on the living organism. One of the possible causes of the effect of hydroxylamine on learning may be hypoxia. Methemoglobin is formed in erythrocytes after the administration of hydroxylamine (Kiese and Plattig 1958). We tried to prevent hypoxia by the injection of methylene blue before the administration of hydroxylamine. This method had already been used by Baxter and Roberts (1959) and is recommended by Goodman and Gilman (1965).

Mice trained in the water maze by the above described method were injected with hydroxylamine or 0.5 M saline 1 hr after the training session. Five min before the injection of hydroxylamine or saline, methylene blue in the dose of 20 mg/kg of body weight (about 0.5 mg per animal) was injected. These mice were tested again 48 hrs after the training session.

Injection of methylene blue, which should prevent hypoxia due to methemoglobinemia, was also ineffective. Trained mice injected with saline had 81.4% retention, mice injected with hydroxylamine had 2.9% retention. Ten pairs of animals were chosen for this experiment. The difference is again highly statistically significant.

Pilot experiments with hydroxylamine showed that this drug has some effect on retention, and this effect is better expressed if the drug is injected after learning.

7.7.1. Effect of hydroxylamine on alimentary conditioning

Some other features of the hydroxylamine action were shown in experiments with alimentary conditioning. In 15 experimental sessions, the conditioned reflex (taking food from the food magazine) was used with mice by the method described above. Four hrs after the last experimental session, the mice were lightly narcotized by ether and the skin of the head cut in the midline. 0.5 N hydroxylamine solution ($HN_2 \cdot OH \cdot HCl$) neutralized with sodium hydroxide to pH 7.4 was administered intracranially by a Hamilton microsyringe. Ten μl of the solution was injected above each hemisphere. The same volume of 0.5 M saline was injected into control subjects. The following groups of animals were tested:

(A) Those injected two weeks before the first experimental session. Fifteen sessions were performed in this series with each animal.

(B) Those injected one week before the first session. Fifteen sessions were performed with each animal.

(C) Those injected one day before the first session. Fifteen sessions were performed with each animal.

(D) Those injected after the fifteenth session, then tested daily for ten successive days starting one day after injection.

(E) Those injected after the fifteenth session and tested daily for ten successive days starting one week after the injection.

(F) Those injected after the fifteenth session and tested daily for ten successive days starting two weeks after the injection.

The results of this experiment are summarized in tables 7.14 and 7.15. The injection of hydroxylamine before the beginning of training (table 7.14) is fully ineffective. We did not find any significant differences between control and hydroxylamine injected animals whose training started one day, one week or two weeks after the injection.

On the contrary, hydroxylamine applied to the trained animals decreases their performance (table 7.15). The longer the period between the injection and first testing session, the greater the decrease of the number of conditioned responses. There are statistically significant differences between the performance in the first session one day after the injection (group D) and two weeks after the injection (group F; $p<0.001$) as well as between the performances of animals one week (group E) and two weeks (group F) after the injection ($p<0.01$). Two weeks after the injection, the performance of animals was almost at the level of untrained mice.

In all the three groups, retraining was possible (see table 7.15).

Control experiments, activity of animals on the rotating rod and in the girdle cage, and the measuring of food consumption showed that the motility and food consumption were slightly lowered 24 hrs after the injection of hydroxylamine, but later these indicators returned to normal.

The meaning of the experiments with hydroxylamine is still uncertain. Only detailed biochemical analysis may show what is actually changed in the brains of hydroxylamine-treated animals. However, our results at least do not contradict the derepressor hypothesis.

Gradual decrease of performance of the animals may indicate that hydroxylamine affected some basic function of the cells of the nervous tissue connected with learning, some process which repeats itself in normal conditions and is gradually and slowly deteriorated as a consequence of a single dose of hydroxylamine.

Therefore, we may again suppose that before the beginning of the training, part of the genome responsible for the production of proteins altering the excitability of certain nerve cells during learning is repressed. Hydroxylamine, which reacts only with active DNA, is therefore without effect. Learning causes activation of this part of the genome, hydroxylamine alters the dere-

TABLE 7.14

Average number of positive conditioned responses after the injection of hydroxylamine before training.

Days after injection	Hydroxy-lamine	Saline	Signifi-cance	Hydroxy-lamine	Saline	Signifi-cance	Hydroxy-lamine	Saline	Signifi-cance
1	0.81	0.81	no						
2	0.62	1.00	no						
3	1.50	1.62	no						
4	2.18	2.44	no						
5	4.18	3.94	no						
6	6.68	5.56	no						
7	7.43	7.75	no	0.92	0.80	no			
8	7.00	7.75	no	0.66	0.50	no			
9	7.06	8.37	no	2.94	2.14	no			
10	8.25	7.75	no	3.22	3.71	no			
11	9.06	8.31	no	3.55	5.71	no			
12	8.68	8.87	no	4.27	4.64	no			
13	8.68	9.19	no	7.22	7.14	no			
14	9.50	9.13	no	7.54	8.50	no	0.98	0.80	no
15	9.18	9.37	no	7.00	8.21	no	3.07	2.60	no
16				9.25	9.57	no	4.14	5.06	no
17				8.33	9.50	no	4.57	5.86	no
18				8.33	9.78	no	6.14	6.13	no
19				8.83	9.35	no	6.43	7.60	no
20				9.11	9.92	no	7.57	7.46	no
21				9.22	10.00	no	7.37	9.13	no
22							9.00	8.46	no
23							9.36	8.80	no
24							9.36	9.53	no
25							9.21	9.60	no
26							9.43	9.73	no
27							9.64	9.93	no
28							9.57	9.00	no
No. of animals	16	16		18	14		14	15	

TABLE 7.15

Average number of positive conditioned responses after the injection of hydroxylamine performed after the 15th experimental session

Day of experiment	Hydroxy-lamine	Saline	Significance (p<)	Hydroxy-lamine	Saline	Significance (p<)	Hydroxy-lamine	Saline	Significance (p<)
13	9.43	9.69		9.25	9.53		9.56	9.75	
14	9.51	9.81		9.68	9.57		9.26	9.33	
* 15	9.62	9.81		9.68	9.92		9.26	9.33	
16	5.68	9.38	0.0001						
17	5.68	8.81	0.001						
18	6.93	9.94	0.0001						
19	8.31	10.00	0.02						
20	7.81	9.88	0.01						
21	7.00	9.50	0.01						
22	6.68	9.75	0.01	3.81	9.28	0.0001			
23	7.00	10.00	0.001	6.37	9.21	0.01			
24	8.00	9.50	no	6.37	10.00	0.01			
25	9.43	9.81	no	6.25	9.14	0.02			
26				7.43	9.64	0.05			
27				8.18	9.42	no			
28				7.81	9.21	0.01			
29				7.68	9.07	0.05	1.53	8.17	0.0001
30				7.75	9.71	no	1.20	8.66	0.0001
31				8.81	9.42	no	2.86	9.00	0.001
32							2.20	9.33	0.0001
33							6.46	9.58	0 01
34							6.26	9.09	0.01
35							8.60	9.58	no
36							9.13	8.92	no
37							9.40	9.67	no
38							9.20	9.67	no
No. of animals	16	16		16	14		15	12	

* Day of the injection of hydroxylamine or saline

pressed DNA and causes misreading of the code (Weigert and Garen 1965). In other words, protein which has been produced is defective, unable to perform some specific function, whether enzymatic, energy supply, alteration of the properties of the postsynaptic membrane, or some other one.

The replacement of the specific active protein by this defective one may be relatively slow. Therefore, the performance of animals tested two weeks after the injection of hydroxylamine is worse than that of animals tested one day after the injection (group D and F). The verification of this hypothesis by a biochemical method will be rather difficult. Derepression or activation of DNA can be proved in vitro, but hardly in the living mammalian brain.

7.7.2. Effect of delayed injection of hydroxylamine on alimentary conditioning

In the following group of experiments, we delayed further the injection of hydroxylamine after the end of the training series. Hydroxylamine was injected intracranially one, two, three, four, six or eight weeks after the end of training, and the series of ten testing trials started always two weeks after the injection. The purpose of this experiment was to learn whether the drug would be effective sometime after the end of training, when the memory trace is not too fresh.

The results shown in table 7.15 demonstrate that intracranial injection of hydroxylamine during the first three weeks after the end of training impairs the performance of animals. The greatest effect appeared after an injection administered two weeks after the end of training. Testing of these animals started four weeks after the end of training. In this group, some animals were not able to be retrained after hydroxylamine injection. On the contrary, hydroxylamine injected four weeks after the end of training (animals tested six weeks after the end of training) is fully ineffective. Hydroxylamine-injected animals are equal to saline controls if injected six and eight weeks after the end of training, too. However both saline- and hydroxylamine-treated animals tested two weeks after these injections, that is eight or ten weeks after the training, show marked gradual impairment of retention.

Our explanation of these facts depends again on the hypothesis that hydroxylamine impairs the retention through a 'mutagenic effect' on activated DNA.

Fifteen sessions of alimentary conditioning was not enough for permanent

TABLE 7.16

Average number of positive conditioned responses after the injection of hydroxylamine in different time intervals after training (12 animals in each group).

	Injection 4 hrs after the last session			1 week			2 weeks			3 weeks			4 weeks			6 weeks			8 weeks		
Session number	hydroxylamine	saline	p	hydroxylamine	saline	p	hydroxylamine	saline	p	hydroxylamine	saline	p	hydroxylamine	saline	p	hydroxylamine	saline	p	hydroxylamine	saline	p
13	9.5	9.8	no	10.0	9.7	no	9.8	9.7	no	9.8	9.8	no	9.8	9.9	no	9.6	9.8	no	9.9	9.9	no
14	9.8	9.3	no	9.9	10.0	no	9.8	9.8	no	9.7	9.7	no	9.8	9.8	no	9.7	9.8	no	9.8	9.8	no
15	9.8	9.3	no	10.0	9.9	no	9.9	9.8	no	9.8	10.0	no	10.0	9.9	no	9.7	9.7	no	9.9	9.9	no
Injection																					
16	1.1	8.2	0.0001	3.0	8.8	0.0001	3.3	9.6	0.0001	5.3	9.7	0.001	9.9	9.2	no	6.1	6.7	no	3.8	2.8	no
17	0.9	8.7	0.0001	2.5	9.0	0.0001	3.3	9.2	0.0001	4.5	9.2	0.001	7.8	7.9	no	5.1	7.4	no	4.2	5.3	no
18	2.8	9.0	0.001	6.9	9.6	0.02	3.1	9.3	0.0001	4.9	9.5	0.001	8.5	8.8	no	6.5	6.9	no	7.2	3.8	no
19	2.4	9.3	0.0001	8.3	9.4	no	3.8	9.5	0.0001	6.2	9.0	no	9.3	9.6	no	6.6	7.4	no	6.7	5.8	no
20	6.5	9.6	0.01	8.2	9.8	no	4.8	9.4	0.001	6.2	9.5	0.02	9.4	8.6	no	6.7	7.8	no	5.5	8.3	no
21	6.7	9.1	0.01	7.9	9.6	no	5.8	9.8	0.001	6.8	9.9	no	8.3	9.8	no	7.7	9.5	no	8.3	8.2	no
22	9.2	9.6	no	9.4	10.0	no	6.2	9.8	0.02	6.1	10.0	0.01	8.8	9.3	no	7.8	9.3	no	7.3	8.5	no
23	9.3	8.9	no	9.1	9.9	no	6.7	9.8	0.02	6.1	9.6	0.02	9.8	9.2	no	8.4	9.2	no	7.3	9.0	no
24	9.2	9.7	no	9.7	9.9	no	6.3	9.7	0.02	6.8	9.7	no	8.8	9.8	no	9.7	9.3	no	7.8	9.3	no
25	9.5	9.7	no	9.6	9.9	no	7.9	9.8	no	7.2	9.5	no	9.3	9.8	no	9.8	9.7	no	7.7	8.7	no

fixation of the memory trace. Some time after the end of the training, the forgetting started.

Our experiments with hydroxylamine may indicate that extinction of the memory trace may be initiated by new repression of the part of the genome activated by learning. Four weeks after the end of learning series, DNA may be already repressed in most animals, and therefore hydroxylamine is ineffective. However, increased levels of RNA and specific proteins are still present in the nerve cells, and therefore the performance of the animals seems to be correct. Later, the level of RNA dependent on learning and level of specific proteins also decreases. This leads to the evident deterioration of the performance of animals, forgetting of the learned task. Hydroxylamine in this last phase is also ineffective, because DNA is repressed.

During the period of testing, we trained the same animals also in another learning situation, passive avoidance. Two weeks after injection of hydroxylamine, on the first day of the testing series of the alimentary conditioning, we trained the mice in a step-through apparatus consisting of a covered wooden chamber to which a narrow lighted chamber was attached. The dark chamber was provided with a grid floor wired to a power supply which administered a 4.5 mA foot shock to the animal when it entered into it.

The mice were placed into a smaller transparent lighted chamber from which they stepped spontaneously into the bigger dark one. The performance of animals was tested again 48 hrs after the training trial. The test trial was conducted in essentially the same fashion as the acquisition trial. The latency of entering the dark compartment was measured with a stopwatch. A 300 sec cut-off time was used for the test trial response latency. The number of animals with good retention, i.e. which did not enter the dark compartment during 300 sec, is summarized in table 7.16. The differences between individual groups were evaluated statistically by a χ^2 test.

We found that the learning ability and retention of passive avoidance was in all observed groups good. It means that even during the time when the retention of alimentary conditioning was impaired by hydroxylamine, the animals were able to learn another task. This may also be the evidence that hydroxylamine affects the areas of the brain engaged in previous learning, leaving areas not yet 'activated' by this learning intact. The animals are then unable to react correctly to the old learned task, but behave normally in new situations.

TABLE 7.17

Passive avoidance in mice two weeks after the injection of hydroxylamine and previous alimentary conditioning; 12 animals in each group. Values expressed in numbers of animals in each group which did not enter the dark space of the step-through apparatus during 5 min of the testing trial.

Interval between the last training session of alimentary conditioning and injection	Saline injected animals	Hydroxylamine injected animals
4 hrs	8	9
1 week	11	10
2 weeks	10	10
3 weeks	10	10
4 weeks	10	9
6 weeks	9	10
8 weeks	11	10
no training	10	10

7.8. *Summary*

In the first parts of the paper, experiments are presented showing that homogenates or extracts of the trained brains affect the performance of naive recipient animals tested in the same training situation. Some limitations of this method are also shown. The animals must starve between the injection and testing session and the donor animals must be trained immediately before the memory transfer. Old memory traces or probably inborn behavioral patterns cannot be transferred. The memory transfer is not accompanied by the incorporation of substances labelled with radioactive phosphorus from the brains of donor animals to the brains of recipients.

In further series of experiments, memory transfer was combined with antibiotics interfering with protein synthesis. Both puromycin and actinomycin D injected to the recipient animals block the memory transfer. Therefore, it is suggested that memory transfer substances act as activators of repressed DNA of nerve cells involved in learning.

Possibility of derepression of DNA during learning was tested also by administration of hydroxylamine. This substance among other effects acts as a mutagen, changing the transcription of DNA templates into RNA and causing the production of defective proteins. It affects derepressed DNA more. We found that it does not interfere with learning if injected before the training series, but it deteriorates the performance of animals if injected after the training. This means that DNA during and after learning may be derepressed. Four weeks after the end of training, hydroxylamine is ineffective again. This may be connected with new inactivation of DNA.

However, it must be admitted that all evidence for the derepression hypothesis presented in this paper is only indirect.

References

ÁDÁM, G. and J. FAISZT, 1967, Nature *216*, 198.

BABICH, F. R., A. L. JACOBSON, S. BUBASH and A. JACOBSON, 1965, Science *149*, 656.

BORENFREUND, E., A. KRIM and A. BENDICH, 1964, J. Nat. Cancer Inst. *32*, 667.

CHAPOUTHIER, G., 1968, Ann. Biol. Clin. *7*, 275.

CORSON, J. A. and H. E. ENESCO, 1966, Psychon. Sci. *5*, 217.

FJERDINGSTAD, E. J., T. NISSEN and H. H. RØIGAARD-PETERSEN, 1965, Scand. J. Psychol. *6*, 1.

FLEXNER, J. B., L. B. FLEXNER and E. STELLAR, 1963, Science *141*, 57.

FREESE, E., E. BAUTZ-FREESE and E. BAUTZ, 1961, J. Mol. Biol. *3*, 133.

FREESE, E. B. and E. FREESE, 1964, Proc. Natl. Acad. Sci. *52*, 1289.

JACOBSON, A. L., F. R. BABICH, S. BUBASH and A. JACOBSON, 1965, Science *150*, 636.

JOHN, E. R., 1967, Mechanisms of memory. Academic Press, New York.

KARLI, P., 1961, J. Physiol. (Paris) *53*, 383.

LUTTGES, M., T. JOHNSON, C. BUCK, J. HOLLAND and J. MCGAUGH, 1966, Science *151*, 834.

MCCONNELL, J. V., 1967, J. Biol. Psychol. *9*, 40.

MIHAILOVIC, L. and B. D. JANKOVIC, 1961, Nature *192*, 665.

MYER, J. S., 1969, J. Comp. Physiol. Psychol. *67*, 46.

REINIŠ, S., 1965, Activ. Nerv. Super. *7*, 167.

REINIŠ, S., 1968, Nature *220*, 177.

REINIŠ, S., 1969, Psychon. Sci. *14*, 44.

REINIŠ, S. and J. KOLOUSEK, 1968, Nature *217*, 680.

REINIŠ, S. and D. R. MOBBS, 1970, Some applications of 'memory transfer' in the study of learning. *In:* W. L. Byrne, ed., Molecular mechanisms of learning and memory. Academic Press, New York, pp. 189–193.

ROSENBLATT, F., J. T. FARROW and S. RHINE, 1966, Proc. Natl. Acad. Sci. *55*, 548 and 787.

ROSENBLATT, F. and R. G. MILLER, 1966, Proc. Natl. Acad. Sci. *55*. 1423 and 1683.

ROSENKRANZ, B. and A. J. BENDICH, 1964, Biochem. Biophys. Acta *87*, 40.

TESSMAN, E. S., 1966, J. Mol. Biol. *17*, 218.

TESSMAN, I., H. ISCHIWA and S. KUMAR, 1965, Science *148*, 507.

UNGAR, G., 1966, Fed. Proc. *25*, 207.

UNGAR, G. and L. N. IRWIN, 1967, Nature *214*, 453.

WEIGERT, M. G. and A. GAREN, 1965, Nature *206*, 992.

CHAPTER 8

Interbrain information transfer: a new approach and some ambiguous data

DAVID KRECH

Psychology Department, University of California, Berkeley, California

EDWARD L. BENNETT

Laboratory of Chemical Biodynamics, Lawrence Radiation Laboratory
University of California, Berkeley, California

8.1. Introduction

A paradigm-crisis (with a bow to T. Kuhn) haunts the field of brain research today. At the very moment when scientific investigators are reporting positive findings in experimental work on interbrain information transfer via chemical agents, other scientists refuse to concede even the possibility of such a transfer phenomenon. Some two years ago we set out to settle this most distressing state of affairs once and for all! Our argument, resplendent in its *chutzpah*, went something like this: We would make an experimental test of the strongest possible case for the weakest possible formulation of the transfer proposition; if even this attempt failed to bring forth convincing positive results, then we, and all other reasonable men, could in good conscience relegate all the interanimal information transfer data to the same dead files where repose ESP records, UFO sightings, proofs of Hullian Theorems and Corollaries, and other such curiosities of science; on the other hand, should our experiments yield robust, replicable positive results, then there would come about a concerted effort on the part of many research workers in attacking and solving what must be the most beguiling of all problems in brain science.

Let us immediately, even if quite unnecessarily, confess that we failed to achieve our brave objective. What we have to report will neither completely confound the opposition nor overly comfort the true believers. Indeed, after completing some 17 experiments involving well over 2,000 experimental animals, all we can report is the Scotch verdict of 'Not Proven'. Our data point to no robust phenomenon, but neither do they clearly deny the existence of *something* out there among the brain macromolecules. Because of this ambiguity, we have held off from publishing until such time as we could clarify the situation. However, because of our continued inability to clear up the ambiguities in our data, and because we thought it desirable to make public all relevant data that might help in the scientific assessment of the transfer phenomenon and, finally, because we believe that our basic experimental design (with as yet untried variations) may merit further trial in other – and fresher – hands, we have decided to wait no longer but publish what we have. In this report, we present neither the procedure nor our data in complete detail, but merely indicate the nature of each. Detailed descriptions of our apparatus, procedure and data are available, on request, to all research workers in the field and may be obtained by writing to us.

8.2. Argument stated

Among the most frequent and compelling of arguments brought forth against the interbrain transfer motion is this: different memories, whatever chemical compounds they may involve, depend upon the differential activation of organized sets of specified anatomical units – neurons, chains of neurons, assemblies of neurons, etc. As long as one believes this, it is difficult to see how a chemical extract from the brain of a trained donor injected via a random route into a naive recipient could possibly activate in the brain of the latter the precise and exquisitely organized spatial–temporal pattern of neural firings which characterizes a specified memory.*

We agreed with this criticism; but, we argued, if we could first prepare the recipient so that the chemical extract from the donor would be injected into a recipient brain that was suitably neurologically differentiated and organized then, presumably, we would maximize the possibility of finding an interbrain information transfer effect. We thought we saw a way to accomplish just that by adapting the Campbell and Jaynes (1966) memory reinstatement procedure (a procedure designed without any reference to interanimal transfer experiments). In their experiment, groups of weanling rats were first trained by being placed in the black and darkened compartment of a two-compartment

* More recently, Ungar (1968) has suggested mechanisms by which injected chemicals may alter and establish functional brain circuits.

apparatus for two 5-min periods, during each of which they received 15 electric foot shocks at programmed but unequally spaced intervals. These shock periods were interspersed by two 5-min no-shock rest periods in the white and lighted compartment. Four weeks later these animals were tested for retention of this infantile experience by being placed in the black compartment which now had an opening into the white one, permitting free access to either compartment. The animal was allowed to remain in the apparatus for 1 hr (with no shock in either compartment), and the time spent in each compartment was recorded during six successive 10-min intervals. The rationale of the test was simply this: if the animal 'remembered' its traumatic experience, then it should prefer the white box; if it had 'forgotten', then the animal's 'natural' preference would take over, and it would prefer the black compartment. Campbell and Jaynes found that the control group, which had had only the original training four weeks prior to the test, did indeed prefer the black box; a second group, which had received the same original training but which had also been given a reinstatement experience every seven days during the 4-week period (each reinstatement consisted of a 1-min 1-shock period in the black box and a 1-min no-shock period in the white box) preferred the white box; and a third group, which had not been given the original training but merely the three reinstatement experiences, behaved like the control group. Thus, original training alone is not adequate to direct a white compartment preference after a 4-week lapse; three weekly 1-min reinstatement experiences alone cannot do so either; but original training plus three reinstatements can change the natural preference of the animal. It should be noted that Campbell showed that this effect could be demonstrated only with weanling animals, as older rats did not need reinstatement to remember (Campbell and Campbell 1962).

We interpreted the Campbell and Jaynes data as follows: a traumatic experience early in life leaves lasting changes in the anatomy and chemistry of the brain. After a sufficient lapse of time, some of these changes decay and, as a result, the entire system reaches a subthreshold level at which it is no longer adequate to control the behavior of the animal. However, the rate of decay can be controlled and the pattern of original morphological–chemical change can be maintained at a suprathreshold level, by minimal 'booster' or reinstatement procedures – procedures which by themselves were inadequate to induce the changes required for effective behavior control. Later, in our work, we added (and tested) the assumption that even if the original pattern of change were allowed to fall below threshold, it could be brought above threshold again by adequate reinstatement procedures.

The relevancy of the above interpretation and hypotheses to the interbrain transfer question may become clearer by indicating how we thought we could adapt the Campbell and Jaynes procedure for our purposes. By first training the recipient animals (a procedure which, to our knowledge, had never been attempted in this area of research), we assumed that we would thereby induce in the recipient's brain the appropriate neural organization (morphological *and* chemical changes) necessary for the display of a preference for the lighted compartment. In the normal course of events, this induced and determining neuronal organization would become non-functional after the lapse of some time, although some elements (and our guess would be that these would be the morphological ones) of this organization would persist. To test whether there was any validity at all to the notion that chemical extracts from one brain can transfer specific 'information' to another brain, we could do one of two things: first, we could attempt to prevent the decay of the originally induced neural pattern in the recipient brain to a subthreshold level by injecting, at various intervals, a brain extract from a freshly trained donor animal. Or, second, after permitting the neural changes in the recipient brain to fall below threshold, we could attempt to 'revive' the pattern and bring it back to a suprathreshold level by the same procedure – by injecting brain extracts from a freshly trained donor. The test of the efficacy of our procedure, in either instance, would be to compare the lighted compartment preferences of our recipient animals with a control group which had had the same original training, was tested after the same interval of elapsed time, but had been injected with brain extracts from untrained donors.

It is perhaps now clear why we considered these procedures as experimental tests of the strongest possible case for the weakest possible formulation of the transfer proposition. We were not attempting to demonstrate the ability of a chemical agent to transfer information de novo from one brain to another;* we were merely examining the possibility that a chemical extract from the brain of a trained donor could act as a booster or energizer in an already existing neural organization in the brain of an identically trained recipient. We might refer to this more modest hypothesis as the 'interbrain memory

* In addition to the experimental groups described in this report, in several experiments we also injected sham-trained rats with trained or naive brain extracts, but no de novo information transfer was observed by the procedures we used.

booster effect'. Clear, replicable, positive support for the interbrain memory booster effect would provide us with two invaluables: (1) data in support of the general hypothesis that different specific 'memories' involve the synthesis. induction, or release of different chemical agents and (2) a reliable behavioral assay method for characterizing and identifying the specific chemical compounds involved in such a memory–chemistry equation.

8.3. Preliminary experiments

Before setting out on the testing of the interbrain memory booster effect, we thought it necessary to carry through two preliminary experiments. The purpose of these experiments was twofold: in the first place, we wanted to reassure ourselves that the Campbell and Jaynes findings could be replicated several times in our hands, with our apparatus, under our laboratory conditions, and with our strain of rats (Holtzman); in the second place, we wanted to check out a possible vitiating factor inherent in our proposed experimental design. This was the injection procedure itself. It should be remembered that irrespective of the material injected, all our animals (control as well as experimental) would be subjected to the painful stimulation and discomfort of fairly massive i.p. injections. Thus, the injection itself could serve as a reminder of the painful original training in the two-box apparatus, and in that way function as a reinstatement experience even for our control group which would be receiving extract from a naive rat.

In our two preliminary experiments we used three basic testing groups (other groups, consisting of very few animals were also run to check out various peripheral matters). These three groups were: a *reinstatement group* (original training and three weekly short reinstatement procedures – à la Campbell and Jaynes – followed by the testing hour one week after the last reinstatement); a *control group* (original training and then testing four weeks later without any intervening reinstatement); and a *NaCl-injection group* (original training plus three weekly injections of NaCl followed by the testing hour one week after the third injection). The first two groups were intended as replicates of the corresponding Campbell and Jaynes groups. Following the Campbell and Jaynes findings, we used as our most reliable differentiating test measure the percentage of time spent in the lighted chamber during the last 10 min of the testing hour.

In the first experiment, the 10 animals of each of the three groups gave the following averages: reinstatement group 91%; control group 37%; NaCl-injection group 41%. In the second experiment the corresponding averages were: 74% (n=9), 41% (n=8), and 15% (n=8). The data from these two experiments were interpreted by us as (1) indicating that we could replicate the original Campbell–Jaynes effect (significant and consistent differences between reinstatement groups and control groups) and (2) that an i.p. NaCl injection would not act as a reinstatement procedure (significant and consistent differences between the reinstatement groups and the NaCl-injection groups, and no significant and consistent difference between the control groups and the NaCl-injection groups).

With these two conclusions reached to our satisfaction, we felt encouraged to proceed with our test for the interbrain memory booster effect.

8.4. Experimental design

In the experiments to be described, four major variants of the experimental design suggested by the reinstatement experiments of Campbell and Jaynes were tried. Three of these are described in this report. (The fourth design departed fairly radically from the Campbell and Jaynes training procedure and is not reported in this paper.) In general terms, the features that characterized all these procedures were training of recipient rats and the trained donors; injection of the recipient rats one or more times with material from trained brains or from naive brains, injection with NaCl, or no injection; and testing.

8.4.1. Training recipients and donors

Male rats were shipped from Holtzman Co. (Madison, Wisconsin) at 21 days of age. The rats were received the next day, weighed, ear tagged and caged three to a plastic cage. Terramycin was supplied in the drinking water for two days. On the third and fourth days after receipt the trained donor rats and recipient rats received their first training. At the time of training the rats were 25 to 26 days of age. The colony was lighted from 7 a.m. to 7 p.m., and training was restricted to 'light' hours. The training box consisted of two compartments, 7 in square and 8 in high, one black and one white (fig. 8.1). The black side had an exposed grid floor; the white side had a white plastic plate over the grid. This training chamber was placed in a large wooden box lined with acoustical tile and with a one-way window in the lid. This provided

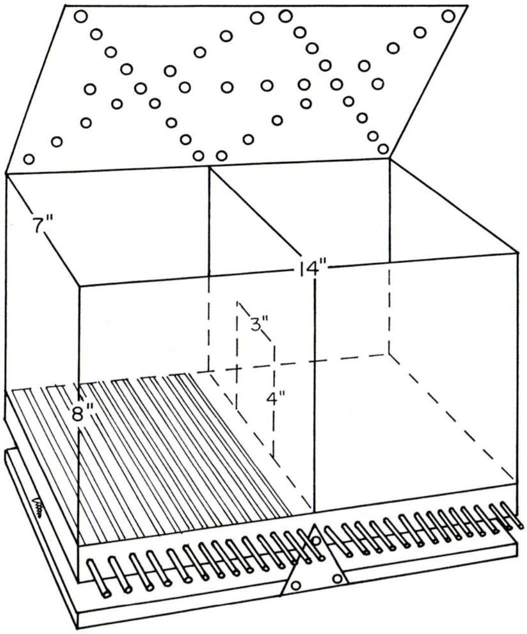

Fig. 8.1. Schematic diagram of training and testing box. The grid compartment was painted black; the other compartment was painted white and had a thin plastic plate over the grid which could be removed for cleaning. During training a solid partition separated the two compartments. For testing this was replaced by a partition with an opening. The hinged lid had a latch (not shown) to prevent the rat from escaping. The center fulcrum allowed the box to tilt slightly when the rat went from one compartment to the other, and this changed the contact with the detecting circuit shown under the left side of the box.

sound damping and permitted one-way observation of the rat by the experimenter. A 7 W bulb in the center top rear lighted both compartments of the training box. A shock duration of 2 sec and an intensity of 1.3 mA from a Grason–Stadler shock generator (E 6070 B) was used during the training, which consisted of a 5-min session with 15 shocks in the black compartment, followed by 5 min with no shocks in the white compartment. This schedule was immediately repeated once, so that one training session required 20 min per rat. Two training boxes were used, enabling us to train two rats simultaneously. Donor reinstatement consisted of one 10-min session – 5 min on the darkened side during which shocks were administered as during the original training session, and 5 min on the lighted side with no shock. In all cases donor rats received the full initial training, followed by reinstatement at weekly intervals. Trained donor rats were reinstated for the last time the day before an 'injection' day. Naive donor rats were never placed in the training apparatus.

8.4.2. Preparation of material for injection

Sacrifice of donor rats typically started about 3:30 p.m. to 4:00 p.m. Animals were sacrificed by decapitation with the Harvard guillotine. The entire brain, including the olfactory bulbs, cerebellum and medulla, was dropped into a Pyrex glass homogenizer filled with 6 ml of cold 0.9% NaCl – 0.01 M tris, pH 7.5. The brains were homogenized two at a time, by means of a Teflon pestle turned by a Mixmaster set at '8'. The glass homogenizing tube was kept in a beaker of ice water to keep the homogenate cold. Homogenates from trained and naive brains were separately pooled in 100 ml graduated cylinders kept in ice. Finally, the homogenates were diluted to a total volume of approximately 5.0 to 6.0 ml/brain. This procedure required about 1 hr for 30–36 rats typically used as donors. Injection was started immediately thereafter. The appropriate volume of whole brain homogenate was injected i.p. into each recipient rat, by use of disposable plastic syringes and disposable needles.

8.4.3. Testing

Prior to testing the rats, the solid partition separating the two compartments was replaced by a partition with an opening 3 in wide and 4 in high. The rat was placed in the appropriate compartment, depending upon the test procedure being used. The apparatus was equipped with electric timer clocks which were actuated whenever the animal was in the darkened compartment. When the rat entered the lighted compartment its weight shifted the apparatus and broke electric contact from the tilt detecting screw on the darkened side.

In the first series of experiments the elapsed time in each box was recorded during the test session. In the other series the times required for the rat to enter the black compartment and remain there for 0.05 consecutive min, and to accumulate total of 1.00 min in the black compartment, were determined. In addition to recording the elapsed time in each box, the number of crossings and the number of boluses dropped were recorded. The order of testing was randomized between experimental groups. All testing was done blind – the technician responsible for the testing had no information about the group assignment. Testing was restricted to the colony 'light' hours, and each animal was tested in the same apparatus in which it had been trained.

In common with most other investigators of memory transfer, we have used the non-parametric Mann–Whitney U test to compare the differences in the test performances of the groups. The p values presented are for a one-tailed test; p values greater than 0.5 indicate that the results obtained were in the reverse direction from those predicted. For those samples for which each N was less than 20, tables of exact significance levels were used (Tec-Ed 1969). When N exceeded 20, Z was computed and normal distribution tables were used. Although more sophisticated statistical analyses may give quantitatively different results, we feel that the overall conclusions will be the same.

8.5. Series 1 experiments

Our first series set the pattern for a whole family of experiments. Three basic groups and methods of treatment were used in experiments 3, 4 and 5: a T group (original training, three weekly injections of whole brain homogenate from a trained and retrained donor group, and testing one week after the last injection); an N group (original training, three weekly injections of whole brain homogenate from a naive, untrained, donor group, and testing one week after the last injection); and a 0 group (original training, no injections of any kind, and testing four weeks after the original training). The amount of brain injected into each T and N recipient progressed from $\frac{2}{3}$ to 1 to $1\frac{1}{3}$ brains during the course of the three weekly injections. The procedure is shown schematically in fig. 8.2.

For testing, the rats were placed in the black compartment, and the proportion of time spent in the white box was determined for six consecutive 10-min intervals. Typically, the proportion of time spent in the white box increased over time. The apparent differences in preferences were greatest during the last testing interval. (Campbell and Jaynes also found this interval to be the most discriminating.) Therefore, for statistical comparisons, the score for the final 10-min interval was used to determine the effects of treatment. The results of the test are summarized in table 8.1. The total number of crossings between compartments and the total number of boluses were also recorded. As no consistent pattern was obtained with these latter measures, no further statistical tests of them were made.

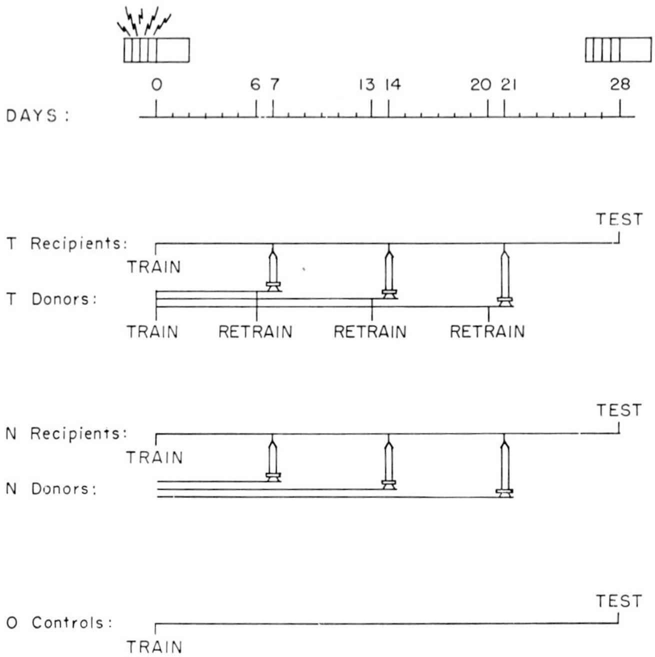

Fig. 8.2. Schematic diagram of the training, injection and testing schedule used in the series 1 experiments (experiments 3, 4 and 5).

8.5.1. Results

In the discussion of results we focus on the difference between the behavior of recipients of trained brains (group T) and the recipients of naive brains (group N).

The first experiment of this series (experiment 3) was encouraging in that the recipients of trained brains (group T) spent on the average nearly twice as much time in the white box as the recipients of naive brains (group N). The difference was not statistically significant, but since it was in the expected direction, additional experiments seemed justified.

The results of experiments 4 and 5 did not confirm those of experiment 3. The N groups in both experiments spent a higher proportion of the last 10-min period in the white box than did the T groups, although in neither experiment was this difference significant. Even before experiments 4 and 5 were in, we considered methods by which the experiments might be improved. Simul-

TABLE 8.1.

Percentage of final 10-min testing interval spent in white box.

Expt. and date	Group injection	No. of recipients	% Time white box	p		
				T versus N	T versus 0	N versus 0
3*	T	8	81	0.12	0.05	
5/67	N	7	45			
	0	11	38			0.35
4*	T	11	50	0.78		
7/67	N	12	64		0.76	
	0	14	66			0.53
5*	T	11	56	0.88		
8/67	N	11	76		0.22	
	0	12	43			0.014
4A**	T	11	39	0.70		
7/67	N	10	46		0.89	
	0	14	66			0.88
5A†	T	9	88	0.23		
8/68	N	9	74		0.043	
	0	10	58			0.11

* Rats in experiments 3, 4 and 5 were given injections one, two and three weeks after the initial training. The amount of brain homogenate progressed from $\frac{2}{3}$ to 1 to $1\frac{1}{3}$ brains/recipient. The rats were tested six or seven days after the last injection by placing them in the black compartment. The percentage of the sixth (final) 10-min test period spent in the white box was determined.

** Rats in experiment 4A were injected with one rat brain two weeks after initial training and were tested two weeks after injection.

† Rats in experiment 5A were injected with one rat brain one week after initial training and tested one week after injection.

taneously with experiments 4 and 5 we ran two experiments of truncated design. The original design had several procedural drawbacks: (1) it required three donor rats for each recipient rat, and particularly when these donor rats were to be trained and reinstated up to three times, the labor required was considerable, and (2) the entire procedure required a month, and we felt that such a lengthy procedure would be a serious drawback for further work, particularly of a biochemical nature.

Therefore experiment 4A eliminated the multiple injection, retaining one injection two weeks after the initial training, with testing two weeks later (fig. 8.3). In experiment 5A we shortened the intervals between training and the injection and between the injection and training to one week. Again, in

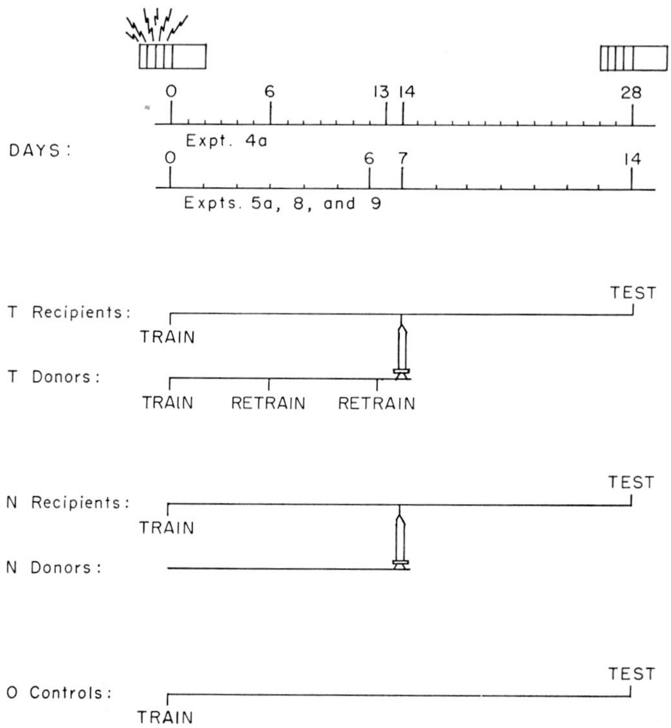

Fig. 8.3. Schematic diagram of the training, injection and testing schedule used in the series 2 experiments. Experiment 4A differed from experiments 5A, 8 and 9 in the time intervals used.

experiments 4A and 5A the results were negative – no significant difference was found between the T and N groups.

Several of the comparisons made between either the T or N groups and the 0 group in this and the subsequent series of experiments do indicate significant differences between brain-injected and non-injected rats. We believe that many non-specific factors are more likely to account for these differences.

These factors most probably would not be related to the problem of information transfer, and therefore we have not pointed them out in our discussion but present them 'for the record' only.

In running these five experiments, several difficulties with the method of testing became apparent. For one thing, the testing method was time consuming: 1 hr was required to test each rat, and even with two testing boxes a maximum of about 18 rats per day could be tested. Perhaps more importantly, some rats immediately 'froze' when placed in the black box. It could be argued that a rat with a good memory of prior events might be expected to 'freeze' in the black box and await the expected shock before attempting to escape. Therefore a high proportion of the time spent in the black box and a correspondingly small proportion in the white box could be interpreted as a result of a *very* good memory rather than as a failure of memory. We thus (perhaps optimistically) interpreted the ambiguous results of the series 1 experiments as a consequence of a fallacy of test method rather than a failure of the experiment. To check out this interpretation, in our subsequent experiments (series 2 and 3) the rats, when tested, were placed in the white compartment rather than in the black, and the time to enter the black compartment was determined.

8.6. Series 2 experiments

In the experiments of series 2 several schedules of injection and testing were tried. These are described in further detail in the footnotes of table 8.2. In series 2, as we have already noted, the rats were placed in the white compartment and the time required for the rat to enter the black compartment was determined. We found that a naive rat, i.e. one that had not been shocked in either compartment, would enter the black side almost immediately. However, a rat which had had the initial shock training and several brief shock reinstatements typically would hesitate a long time before entering the black box, and sometimes failed to do so within an hour. In the second series of experiments two measures were taken: (1) the time required for the rat to enter the black compartment and remain there for 0.05 min ($t_{0.05\ min}$); and (2) the time required for the rat to reach a cumulative total of 1.00 min in the black compartment. Since the $t_{1.00}$ appeared to give no better indication of a treatment effect than the $t_{0.05}$, only the $t_{0.05}$ results are presented in table 8.2.

As soon as a rat reached a cumulative total of 1.00 min, it was removed

TABLE 8.2

Time required for recipient rats to enter black chamber for 0.05 min ($t_{0.05}$)..

Expt. and date	Group injection	No. of recipients	Initial test				Retest			
			$t_{0.05}$ (av.)	p: T versus N	p: T versus 0 (or NaCl)	p: N versus 0 (or NaCl)	$t_{0.05}$	p: T versus N	p: T versus 0	p: N versus 0
6*	T	10	20.6	0.13			4.6	0.075		
11/67	N	12	13.9		0.15		3.0		0.014	
	0	12	15.5			0.34	2.3			0.20
7*	T	23	9.3	0.22			2.7	0.74		
12/67	N	18	5.0		0.10		3.0		0.51	
	0	24	4.8			0.38	2.3			0.21
8**	T	25	37.1	0.08			7.8	0.93		
12/67	N	24	27.9		0.006		16.1		0.08	
	0	23	21.3			0.09	8.6			0.008
9†	T	9	31.1	0.55						
1/68	N	5	35.7		0.46					
	0	12	30.7			0.50				
9A†	T	11	44.0	0.27						
1/68	N	9	37.9		0.014					
	0	12	20.0			0.05				
10A††	T	12	28.9	0.20						
2/68	N	12	25.7		0.27					
	NaCl	12	29.1			0.50				
10B††	T	12	4.1	0.93						
3/68	N	12	6.8		0.28					
	NaCl	12	2.2			0.015				

* Rats in experiments 6 and 7 were given injections one, two and three weeks after initial training and were tested one week after the last injection, i.e. four weeks after initial training. The retests for experiments 6 and 7 were done six weeks after the initial training.

** Rats in experiment 8 were given only one injection of brain seven days after initial training, tested one week later, and retested three weeks later.

† Rats in experiment 9 were given one injection 11 days after training, were tested six or seven days later. Rats in experiment 9A were given one injection 11 days after initial training and were tested one or two days later.

†† Rats in experiment 10A and B were given one injection 12 days after initial training and were tested one day later. Experiment 10B was done one week later than experiment 10A.

from the test chamber. The average $t_{1.00}$ was close to 30 min. This permitted us to test more rats per day than was possible in the procedure of series 1. A rat not reaching the $t_{1.00}$ criterion by 60 min was also removed and given a $t_{1.00}$ score of 60 min. Some rats never entered or remained in the black compartment for 0.05 min. Such rats were scored as 60 min for both $t_{0.05}$ and $t_{1.00}$.

In experiments 6, 7 and 8 the rats were retested one to three weeks after the first test without any intervening treatment (injection).

In the first of this series (experiment 6), group T had a longer latency than group N, with a p value of 0.13. In the retest, group T again had a longer latency than group N, and the p value reached 0.08. In the second and larger experiment (experiment 7), while the trends observed in the initial test of experiment 6 were confirmed, the p values obtained were less significant.

About the same time that experiment 7 was being run, we also ran a hybrid variant of experiment 4A of series 1 and experiments 6 and 7 of series 2. In this variant (experiment 8), each rat was given only one injection of a brain, seven days after initial training, followed by tests one week and three weeks later. Again, an encouraging difference for the mean value of $t_{0.05}$ (37.1 min for T versus 27.9 min for N) was obtained between the two brain-injected groups with a p value of 0.08. On retest, group N had the highest latency.

The design of experiment 9 was similar to that of experiment 8. Experiment 9 yielded no significant differences between the two recipient groups. However, it should be pointed out that this experiment was characterized – for unknown reasons – by an atypically high mortality rate. Ten of the 24 recipient animals died. In previous (or in any of the subsequent) experiments, we rarely observed a mortality rate as high as 10%.

Experiments 9A, 10A and 10B were further truncated versions of experiments 8 and 9 (see footnote to table 8.2). It should be noted that for experiments 10A and 10B (and for all subsequent experiments), the 0 group became a NaCl group, i.e. the rats were injected with an equivalent volume of 0.9% NaCl. The results of experiments 9A and 10A again suggested that although group T and N did not differ significantly, the difference was in the expected direction. For unknown reasons, the latencies for all the groups were unexpectedly low in experiment 10B, and the differences between the T and N groups were in the direction opposite to that predicted.

At this time we assessed all the data of series 2. Five out of seven experiments had yielded mean differences between T and N groups in the expected direction. The two experiments which showed a reversal (9 and 10B) could be 'explained away' – thus, in experiment 9 there was an unusually high mortality rate among the injected animals, and in 10B there was an atypically short latency which we now believe may have been due to malfunctioning of the shock apparatus during training. (However, we have only circumstantial evidence for this belief.)

The average p values for the seven experiments were far better than those observed for the five experiments in series 1. All told, it appeared to us that in series 2 we had gained some ground.

In evaluating series 2, it appeared that of all the variations tried, the one employed in experiment 8 (one injection with testing one week later) was the most promising. With this in mind, we embarked on our series 3 experiments.

8.7. Series 3 experiments

The experimental design used in experiments 14 through 17 were replications of experiment 8 through the first testing period but added a new 'twist' to the retests, that is, additional injections prior to each retest (fig. 8.4). The variant in the retest derived from the following considerations: (1) one of the effects of the first testing procedure for both the T and N groups could be partial extinction of the fear of the black compartment, since in this testing procedure entry into the black box did not result in a foot shock. If such extinction did take place, this would explain why, on the second test, no consistent differences between the T and N groups were observed. And (2) if additional trained-brain injections were made prior to the retests, the fear of the black compartment could again be reinstated. Now one might even expect larger differences between T and N groups on the retest than on the original test.

The results of this series (summarized in table 8.3), when viewed as replications of experiment 8, were disappointing. In only one experiment (experiment 15) were strong positive results found on the first test. The T group showed a higher latency than the N group, and this difference was significant at the 0.003 level. However, in the other three experiments (14, 16 and 17), the differences were either non-existent or in the 'wrong' direction, and in no instance was the p value significant.

This series presents a different picture when we examine the results of the second test, which had now been preceded by another injection. For this

TABLE 8.3

Time required for recipient rats to enter black chamber for 0.05 min ($t_{0.05}$).

Expt. and date	Group injection	No. of recipients	Test after first injection: $t_{0.05}$ (av.)	p: T versus N	p: T versus NaCl	p: N versus NaCl	Test after second injection: $t_{0.05}$	p: T versus N	p: T versus NaCl	p: N versus NaCl	Test after third injection: $t_{0.05}$	p: T versus N	p: T versus NaCl	p: N versus NaCl
14*	T	15	10.8	0.72			8.5	0.28			2.0	0.33		
8/68	N	15	9.2		0.17		3.6		0.034		2.0		0.22	
	NaCl	15	6.1			0.07	2.2			0.10	2.9			0.50
15*	T	16	16.6	0.003			11.1	0.31			6.7	0.57		
9/68	N	14	8.8		0.09		4.9		0.03		2.9		0.08	
	NaCl	16	13.4			0.96	1.2			0.17	0.7			0.07
16*	T	21	18.6	0.85			10.7	0.40						
10/68	N	23	24.6		0.86		6.0		0.24					
	NaCl	26	22.6			0.49	6.2			0.31				
17*	T	16	39.0	0.57			8.7	0.92						
3/69	N	17	40.2		0.33		22.4		0.23					
	NaCl	16	33.6			0.29	11.1			0.07				

* In experiments 14, 15, 16 and 17 the rats were injected with one brain/recipient one week after initial training. The recipients were then tested one week later. After an additional six days the rats were reinjected with one brain/recipient and tested one day later, i.e. three weeks after initial training. In experiments 14 and 15 the rats one week later received a third injection of one brain/recipient and were tested for the third time one day later.

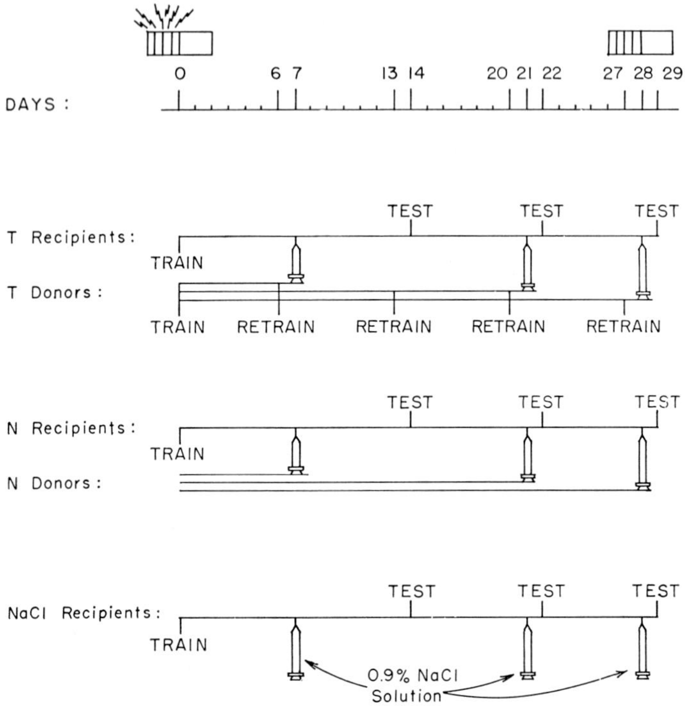

Fig. 8.4. Schematic diagram of the training, injection and testing schedule used in the series 3 experiments (experiments 14 through 17).

second test, experiments 14, 15 and 16 yielded differences which were in the expected direction but, again, no one of these differences reached an acceptable p value. An analysis of each animal's performance on tests 1 and 2 showed that several animals with very low latencies on test 1 gave extremely high latencies (20 to 60 min) on test 2. This was true for about 15% of the T animals and for none of the N animals. These findings are congruent with the hypothesis that trained-brain injections, made prior to the retest, could reinstate the extinguished fear of the black compartment. However, the final experiment of this series (17) failed to confirm any of the findings of the preceding three experiments.

It would appear, in summary, that in series 3 we find the characteristic refrain of so much of the work in memory transfer; the experiments fail to replicate the 'encouraging' results of previous work, but at the very moment of this failure, they open new and 'promising' avenues and beckon the experimenter on for yet still another try.

8.8. Discussion

Almost all of the interbrain information transfer work is addressed to one of two questions (or sometimes both questions simultaneously): does the phenomenon exist? and what is the identity of the active principle(s)? It seems to us, however, that the first question has priority. Unless it can be answered affirmatively, the second question, quite obviously, is without meaning. And to answer the first question affirmatively, we must discover an experimental procedure which will yield consistently positive results in the hands of all qualified experimenters and in different laboratories. Weak positive answers, only fitfully obtained, and only among a chosen few experimenters, will not make for much progress in this entire enterprise. However, once we can be certain that the phenomenon exists, and that it can be observed at will, then and only then will we have prepared the way for substantial progress on the second great question of this beguiling phenomenon. For, when such a stage is reached, the problem of isolating and characterizing the active principles involved in interbrain memory transfer (or even 'memory boosting') will undoubtedly arouse the excitement and enlist the sustained efforts of many different scientists from a wide variety of disciplines. And it is our belief that the answer to this second question (which may indeed become *the* great question of biology) will be a multidimensional one.

One of us (E.L.B.) has had a long-standing interest in this field and has attempted, repeatedly, to test the phenomenon of interbrain memory transfer via several of the procedures described elsewhere in this volume (especially those of Ungar and Byrne). While in some few of these experiments statistically significant differences in the expected direction have been found between the experimental groups, the differences have not been consistently replicated from experiment to experiment. The undependability of the behavioral findings has not only made serious biochemical work unfeasible, it has also made questionable whether the phenomenon exists at all. As of this writing, we believe this scepticism characterizes the thinking of many of the scientists who have interested themselves at all in this problem.

It was with this assessment of the state of the art and this background of frustration and uncertainty that the two of us initiated the experiments described in this paper. As stated in our introductory sections, we believed that a good case could be made for the notion that a prepared brain might be more receptive for the memory molecules than the unprepared or naive brain of the recipient typically used in transfer experiments. Our results did not live up to our expectations. Although an occasional significant difference in the expected direction was obtained between the trained and naive recipient groups, we were not able to reproduce these findings in a satisfactorily consistent manner.

The meaning of our 17 experiments, however, is not truly told by the testimony of each experiment taken by itself. Appropriate statistical tests, applied to all the experiments taken together, or to the combined experiments belonging to any one experimental design, might indicate a highly significant difference between the performance of trained-brain and untrained-brain recipients. Such a statistical finding might even provide reasonable support for evidence in the interbrain information transfer phenomenon. Indeed, Professor Jerzy Neyman has been interested in the development of just such statistical tests for our data. These tests and the resulting analyses may prove to be sufficiently interesting to warrant presentation in the future. It is quite possible that the addition of more data – similar in kind to those presented here – would establish beyond reasonable doubt that the phenomenon does exist. On the other hand, it is reasonably clear that even if we should demonstrate the existence of such a phenomenon, none of the experimental procedures we have thus far tried can yield the kind of dependable and replicable results which the biochemist would desire (and need?) in order to isolate the active principle in brain. Here, then, we are presented with questions of research strategy and tactics. It would seem to us that any future work should be directed not toward the accumulation of additional data by any one of the procedures we have described, but rather toward obtaining a procedure that can yield more consistent and robust findings.

The possibilities for such future work are almost endless. Within the basic theoretical rationale and experimental designs described in this paper, there remain many untried variations which we believe merit serious consideration. There are two major sets of variables upon which major changes might with profit be played: the behavioral and the biochemical. Thus, on the behavioral side, relatively little imagination is required to suggest different training and testing designs, and training, injection and testing schedules that might be more successful. Of especial interest may be the suggestion that injections follow the training and extinction of the original habit and that the testing be a test of postextinction reinstatement. This would have the added advantage that all the animals, the trained-brain and naive-brain recipients, would be brought to the same performance level prior to their injection and testing.

On the biochemical side, the most obvious changes to explore would be in methods of preparing the homogenate. Throughout all our experiments, we prepared our injection material from whole brain. It was our belief that if there are memory molecules within the brain, we would be most likely to detect them by using an extract of whole brain. However, it is possible that these molecules may be located in discrete areas of the brain, and their potential effects are damped or antagonized by molecules in other areas of the brain. It may, therefore, be worthwhile to test homogenates from different areas of the brain taken one at a time. But quite aside from where we seek our active brain material, there is the question of what we extract. Our feeling was that we would be most likely to find the active material within the total homogenate rather than in some fraction, such as a crude nuclear precipitate or brain supernatant and that, when and if the phenomenon could be established with the undifferentiated preparations, we could then more readily proceed to find the best way to isolate and purify the active material. But here again it is obvious that any undifferentiated preparation may have contained too much inert material to make possible an adequately large dose of the active material. In other words, it may be very worthwhile to attempt various extracts, rather than depend solely on total homogenates, even in the initial experiments.

At this point we prefer to reflect further on the results of our already completed experiments before embarking in new searches for the end of the rainbow. We were prompted to write this report, as we indicated at the beginning of this paper, by the hope that other, yet unbloodied, heads might be induced to explore some of these new avenues. If such there be, we wish them well – indeed, we may meet them on the way.

References

CAMPBELL, B. A. and E. H. CAMPBELL, 1962, J. Comp. Physiol. Psychol. *55*, 1.

CAMPBELL, B. A. and J. JAYNES, 1966, Psychol. Rev. *73*, 478.

TEC-ED, P.O. Box 337, Berkeley, California: Exact significance levels of the Mann–Whitney two sample statistics.

UNGAR, G., 1968, Perspectives Biol. Med. *11*, 217.

CHAPTER 9

An apparent transfer effect in chickens fed brain homogenates from donors trained in a detour task

SHELDON B. SPARBER and EUGENE ROSENTHAL
Department of Pharmacology, University of Minnesota
Minneapolis, Minnesota 55455

9.1. Introduction

Biochemical changes occur during and after the learning experience (Hydén and Egyházi 1962; Hydén 1963; Hydén and Lange 1965). Interference with synthesis, release, receptor attachment or metabolism of several endogenous substances have all been shown to affect performance in either direction (Whitehouse 1964; Flexner et al. 1964, 1966; Berger and Stein 1969). Because of early experiments with nucleic acids (Cook et al. 1963; Brown 1966; Solyom et al. 1966; Wagner et al. 1966; Siegel 1967) and ribonuclease (Corning and John 1961) much effort has been expended in trying to delineate the role these substances play in integration, storage and retrieval of information. It therefore seems obvious that studies into the nature of memory transfer should also concentrate upon the role of RNA or DNA as the substrate responsible for the transfer effect. Those who dispute the importance of the nucleic acid compounds have often supplied alternatives (e.g. proteins or polypeptides, see Ungar and Oceguera-Navarro 1965). More importantly, the inability of several laboratories to replicate some of the positive findings (Gross and Carey 1965; Byrne et al. 1966; Gordon et al. 1966; Luttges et al. 1966) has prompted an attitude of disbelief that the phenomenon of 'memory transfer' exists at all. The lack of reproducibility has been explained in many ways, ranging from animal (strain, etc.) differences to biochemical extraction procedural differences. Our approach to the study of memory or learning transfer incorporated part of the experimental design that was originally used by McConnell (1962) in his study with planaria. The cannibalistic nature of the study would allow circumvention of the methodological problems inherent in extraction and chemical identification experiments. If transfer could be demonstrated, the choice of the chicken, a phylogenetically more advanced species, would have several consequences. Among these are included the indisputable demonstration of whether or not learning occurred in the first place, an issue still in dispute regarding the capabilities of planaria (McConnell 1966). In addition, the experimental model we chose enabled us to use a very young but precocious subject, whose advantage of a short behavioral history obviated having to deal with the infinite variables resulting from age and exposure to postnatal life. The rat is a case in point. By the time it is old enough to be used in behavioral experiments it will have been alive for at least 30–50 days or longer. To identify, after transfer, one behavioral response from a repertoire of perhaps thousands would require more than most of the scientific community is willing to admit. In some experiments on chemical transfer of information, the intracranial route of administration of brain homogenates or extracts was used. In others, substances were injected i.p. Much of the criticism directed at the transfer studies which implied a role for nucleic acids stems from data which show little or no ^{32}P-labeled RNA in the brains of rats injected with this substance (Luttges et al. 1966). If RNA, or any other fairly large molecule is directly involved, it might have greater accessibility to the brain of the chick, with its incompletely developed blood–brain barrier (Waelsch 1954; Bakay 1955; Lajtha 1957). In addition, the entire gastro-intestinal tract of the chicken has an acid pH decreasing the possibility of dissolution and/or destruction by the basic environment found in the intestine of most other species.

Perhaps, equally as important as the aforementioned reasons for the choice of the chicken is the comparative nature of the experiments. If transfer of information could be demonstrated in another vertebrate class and species it should add further evidence that the phenomenon is real. Greater generality in turn might be conducive to more widespread interest in this important area. Ironically, the need for continued research was amply supported by the combined statement of several laboratories that had been unsuccessful in attempting to replicate the transfer experiments which had yielded positive results with rats (Byrne et al. 1966).

There have been many suggestions that transfer effects purported to have occurred are nothing more than estimations of movements which lack specificity and therefore objective quantitation. 'Umweg' or detour learning (Kimble 1961) in the chick neonate has previously been used in studies of neurochemical–behavioral development (Scholes 1965) and behavioral teratology (Sparber and Shideman 1969). In detour behavior the learned task seems to be imcompatible with innate behavior which is displayed upon initial exposure to the situation. The incompatibility of the innate response with that of the required detour response seemed to offer some advantages in studying specificity.

The experiments with detour learning which follow, include an attempt to establish simple, standard procedures. After establishing that transfer of a detour response could be demonstrated, we proceeded to a more complex behavioral task to determine if the transfer effect is position specific. Although much time and effort has been expended, no clear-cut indication of position specific transfer has appeared. These attempts will be described and all data will be presented in different forms so that the reader can choose to interpret them in his own way.

9.2. Experiments

9.2.1. Experiment 1

Before the transfer experiments were carried out we examined some of the variables controlling detour behavior. As a result of that preliminary work, we chose to train the donor chickens with a massed trial design and test for transfer with a spaced trial design. Because the difference in design might somehow affect the outcome of the experiment, due to a type of 'state dependency', we performed the first experiment. Our intention was to compare the learning curves of the two groups (massed versus spaced) given the same number of trials with equal exposure time.

Apparatus. The detour apparatus is a simple box with a plexiglass wall isolating the experimental subject from food, broodmates and a light, which are provided as goal objects and collectively termed the social side of the apparatus (fig. 9.1). Chicks on the social side acted as 'stooges' and were never used as subjects in the experiment. An opaque metal tunnel provided passageway between the two sections. For observation purposes, a two-way mirror served as the wall opposite the tunnel on the social side of the apparatus. During experimental sessions the room lights were turned off to accomodate the mirror. In addition, a masking noise (white noise) was present so that extraneous noises would have few consequences.

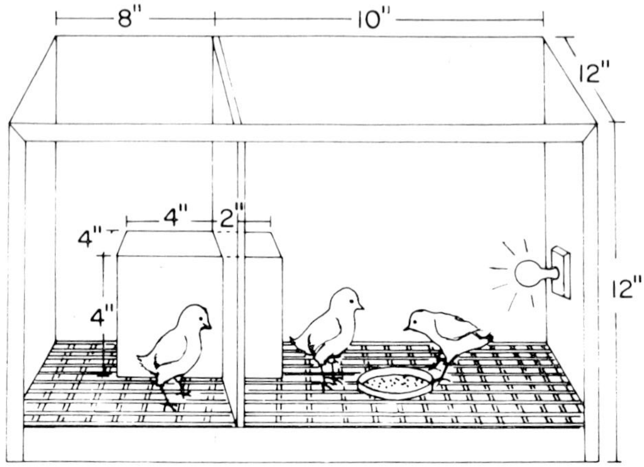

Fig. 9.1. Single tunnel apparatus used to study detour learning and transfer in the young chick.

Procedure. A food-deprived white Leghorn chick, hatched in our laboratory, (deprived for 12 hrs starting the night before each trial) is placed in the center of the isolated section of the apparatus. The chick now has a choice between two incompatible behavioral responses: (1) the immediate response of the hungry subject in isolation is to push vigorously along the plexiglass partition trying to gain access to the goal objects, (2) alternatively, the chick can turn away from the partition and go through the tunnel (detouring). If a chick does not detour through the tunnel within 4 min, it is guided through by the experimenter with a wooden probe. Once on the social side of the apparatus the chick is allowed to remain there for 30 sec and then removed, terminating a trial. The response latency in seconds is measured from insertion of the subject into the isolated portion of the apparatus to emergence of the subject from the tunnel into the social side. At the end of an experimental session the chicks are fed again.

In this experiment, chicks were randomly assigned to each group and given either massed or spaced trials. All learning trials were given at the same time of the day. Chicks were three days of age at the beginning of training. The spaced-trials group was given one 4-min trial per day for seven days, thus receiving six consecutive trials from the third day of age through the eighth with one retention trial given on the eleventh day of age. Massed trials were given to another group of chicks starting at seven days of age. This training consisted of five consecutive 4-min trials, with a sixth 4-min trial at eight days of age, and a retention trial at 11 days of age. A third group, broodmates to the other groups, was kept as controls for the transfer experiments.

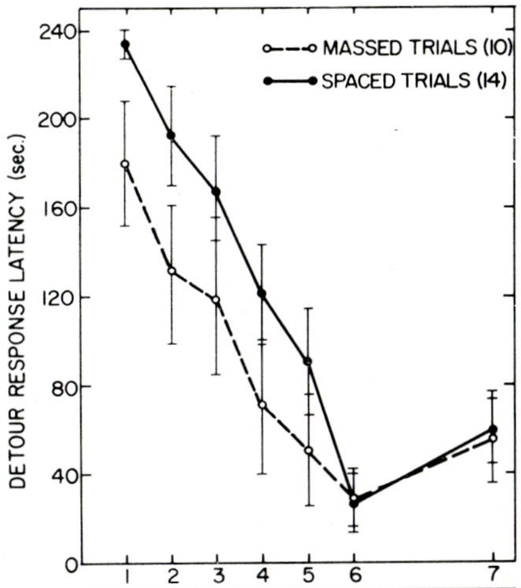

TRIAL	1	2	3	4	5	6	7
AGE (DAYS POST HATCH)							
massed	7	7	7	7	7	8	11
spaced	3	4	5	6	7	8	11
RESPONSES<4 MIN.							
massed	6	6	8	9	9	10	10
spaced	1	4	7	12	11	14	14

Fig. 9.2. Massed and spaced trial detour learning in the chick. Points are group means with vertical lines representing the standard errors of the means (SE). Values in parentheses are the number of subjects in each group.

Results. As seen in fig. 9.2, some chicks have difficulty in learning the detour response at a very young age, as previously reported by Scholes (1965). However, by the third trial, 80% of the massed group (seven days of age) and 50% of the spaced group (five days of age) responded in less than 4 min. On the following trial, 90% of the massed and 85% of the spaced group made the detour response. By the fifth and sixth trial, asymptotic levels of mean response latencies were reached. After the seventh trial, at 11 days of age, the brains of the group given massed trials were removed, pooled and homogenized (glass tube fitted with teflon pestle, Tri-R Instruments, N.Y.) in one-half vol of cold 0.9% NaCl solution. This homogenate was used for the transfer experiment.

Discussion. Scholes (1965), using a learning criterion and a massed-trials procedure, reported that chicks learn to detour when at least seven days of age. However, examination of individual latencies shows that learning occurs as early as the initial exposure to the situation at three days of age. Use of a spaced-trials design with one trial per day enabled this observation.

Detour learning seems to contain numerous operant components. Many of these operants need further study before complete stimulus control can be maintained. The size of the apparatus and visual cues within the apparatus, number of broodmates on the social side and deprivation schedules, all seem to affect response latencies and topographies. These variables, among others, can alter the general developmental pattern of this response under otherwise identical testing conditions. In addition, a chick's exact age in hours, from the time of day it hatched, can account for large differences in initial exploratory activity in a novel environment. At approximately three to three and one-half days of age a naive animal placed in the detour apparatus initially freezes for an extended period of time. If, however, the animal is less than three days of age, exploratory activity can give rise to spontaneous detour responses, shifting the learning curve to the left (see experiment 2). Perhaps the most important information resulting from the preliminary and developmental experiments is the importance of maintaining adequate controls. We suggest that entire experiments be carried out under conditions which will diminish the variability resulting from the use of subjects from different broods (different litters in rodents) hatched at different times of the year, and reared under so called identical conditions, which can never be really identical. If the dynamics of development did not interact so intensely with acquisition of a learned task, control for this type of variability might not be such a necessity.

9.2.2. Experiment 2

Having determined the validity of a spaced one trial/day schedule as a measure for acquisition of the detour response, we proceeded with the first transfer experiment.

Procedure. In order to avoid possible changes resulting from storage of brains or homogenates, we planned to have newly hatched chicks available on the day of sacrifice of the donors. Chicks of the one-day-old brood were randomly assigned to three groups: experimental, control and a third group to act as stooges. One-half ml of trained donor brain homogenate was fed to chicks of the experimental group using a tuberculin syringe and a blunted 22 gauge needle (fig. 9.3). The control group was fed an equal volume of brain homogenate from the broodmates of the chicks used in experiment 1. These control donors had never been exposed to the detour apparatus. Two

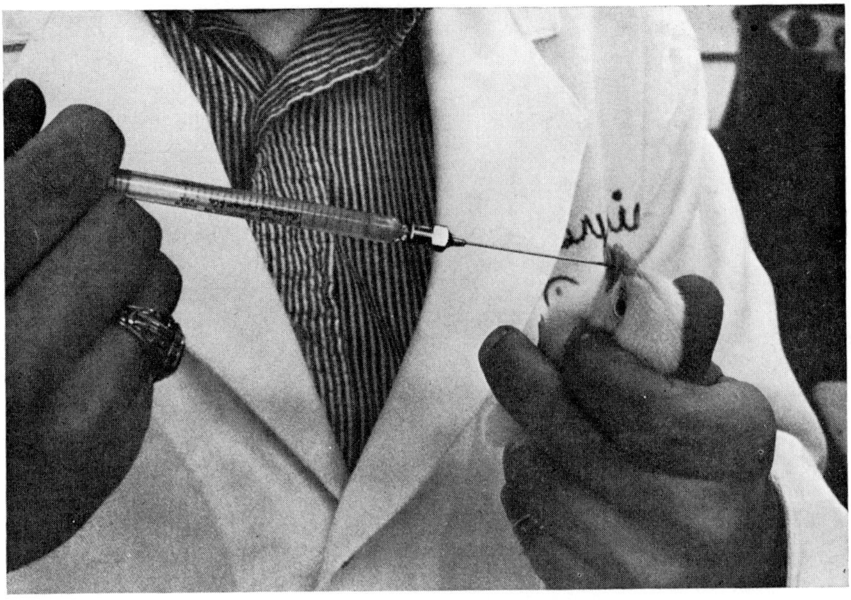

Fig. 9.3. Feeding brain homogenate to newly hatched chicks. Most chicks readily take the homogenate once they taste it.

days later (so that the possibility of a pharmacologic sensitization or stimulation effect is minimized), at three days of age, both groups were presented with the detour situation and given one trial per day until they were eight days of age.

Results. The animals that received brain homogenate from experimental donors performed much better than controls (fig. 9.4). At the second trial, for example, one out of ten control chicks and nine out of ten experimental subjects responded within 4 min. By comparing the means of individual cumulative response latencies throughout the experiment, it can be seen that the greatest differences appear in the early trials, this difference becoming smaller as the control group learns to detour. Still skeptics regarding the reality of these differences, we decided to feed the remainder of the homogenate (frozen and kept over solid carbon dioxide since the preparation) to the next brood of chicks due to hatch within two weeks.

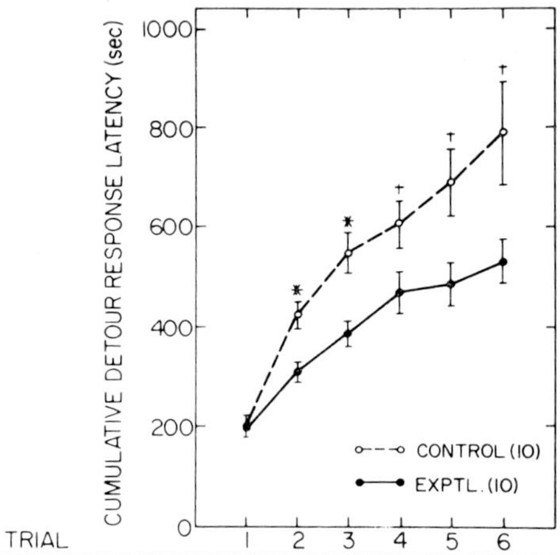

AGE (DAYS POSTHATCH)	3	4	5	6	7	8
RESPONSES < 4 min						
CONTROL	4	1	8	9	8	8
EXPTL.	5	9	9	9	10	10

Fig. 9.4. Detour response latencies of the first transfer experiment. Means (± SE) were derived from individual cumulative performance records. * $p < 0.0025$, † $p < 0.025$, ‡ $p < 0.05$ (single-tail t test).

The seond transfer experiment was run in a manner identical to the first, except this time a blind procedure was followed, SBS feeding the homogenate to the chicks and ER running them in the detour apparatus. Although bias in the first transfer experiment could hardly account for the great differences observed, this additional control seemed necessary for our satisfaction. Essentially, the information derived from the second experiment supports the original observations. Again, the group receiving brain homogenate from chicks having learned the task had significantly better performance records (fig. 9.5).

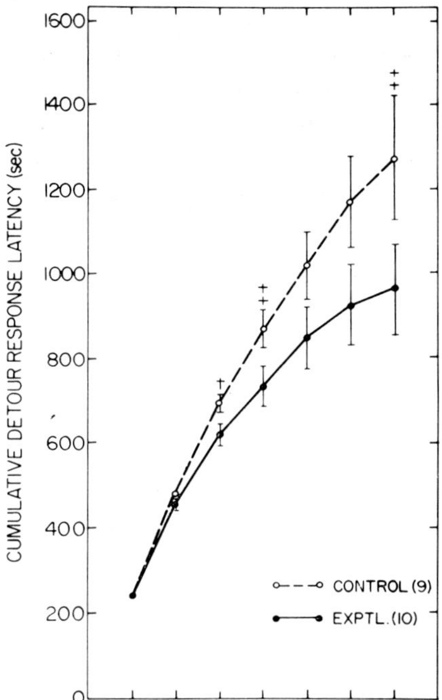

TRIAL	1	2	3	4	5	6	7
AGE (DAYS POSTHATCH)	3	4	5	6	7	8	9
RESPONSES < 4 min CONTROL	1	1	3	5	6	5	7
EXPTL.	0	2	7	7	8	9	10

Fig. 9.5. Detour response latencies of the second transfer experiment in the single tunnel apparatus. See legend to fig. 9.4 for explanation of symbols.

Discussion. A comment seems warranted regarding the apparent difference in early performance of both control and experimental groups in the first and second transfer experiments. In our first transfer experiment, the chicks hatched during the latter part of the day and through the night; thus by the third day of age they were, in essence, two and one-half days old. In the second transfer experiment the chicks started to hatch early in the day and consequently at three days of age they were really more like three and one-half days old. This difference in age could account for the differences in the first and second exposure in both transfer experiments. If one assumes that in the second transfer experiment (fig. 9.5) the second trial (at four days of age) was essentially equivalent to the first trial in fig. 9.4, the curves are quite similar in shape and value; i.e. although absolute cumulative response latencies vary from experiment to experiment, the order of the magnitude of the differences between experimental and control groups is about the same. In addition, it must be remembered that freshly extirpated and homogenized brain was used in the first experiment while in the second transfer experiment the remainder of this homogenate was used after being frozen and stored for about three weeks. It is possible that this procedure resulted in an alteration of whatever principles are responsible for the performance facilitation (e.g. oxidation or enzymatic attack while being thawed etc.).

9.2.3. Experiment 3

Although we were reasonably certain in our interpretation of the data in terms of a real transfer effect with some semblance of specificity because of the topography of the response, we became engaged in a series of new experiments which might support these findings. It entailed the use of a larger apparatus containing two tunnels, one on the near and one on the far wall, both going through the plexiglass partition (fig. 9.6). We intended to train young chicks to go through either the mirror tunnel no. 1 or the other, no. 2, until reaching asymptotic latencies. When their brains, or the stooge brains, were fed to recipient chicks we hypothesized: (a) facilitated performance by recipients of trained brain and (b) preference for the tunnel to which the donors were trained.

Because the walls and tunnels of the newer apparatus were fabricated from aluminum sheet metal it was difficult to visually discriminate between tunnel and wall. We remedied this condition by painting the outside of the tunnels black. The subsequent enhancement of responding resulted in performance similar to that described for the single-tunnel experiments. Having established the feasibility of using the more complex apparatus, we proceeded to train all donors for the position-specific experiment to detour through a given tunnel. We then planned to divide this group into two subgroups. One subgroup was to receive further training with the original tunnel blocked, thus reversing them to the opposite tunnel. We expected to see either a transfer of a position-specific detour response, or if the response itself is too general in nature to show position specificity. We expected it would be easier to reverse the group of naive birds fed homogenate from chicks having been reversed.

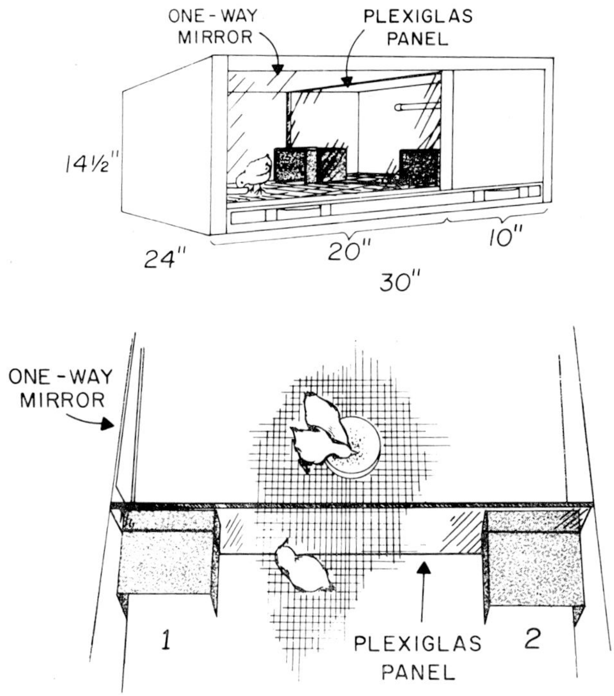

Fig. 9.6. Two-tunnel apparatus used in position-specific transfer experiments.

Procedure. Donor chicks for all experiments in the two-tunnel box were trained, by various schedules, until responding to the appropriate tunnel reached asymptotic levels. When training the donors, a removable plexiglass wall was inserted in slits in either tunnel to block the inappropriate response (detour through the wrong tunnel). With the removable wall in place the bird could enter but not go completely through. The experiment to be described at this point used four donor and four recipient groups. The trained donors were given enough 2 min guided trials to the appropriate (no. 1 or no. 2) tunnel so that terminal detour latencies averaged 3 sec. The non-trained donors consisted of a group of stooges and a group of broodmates kept in their home brooders and never exposed to the apparatus.

All donors were sacrificed at the same time and their brains homogenized separately in a volume of 0.9% NaCl which was the difference between 3 ml and a volume equal to the weight of the brain, e.g. if the brain weighed 1 g it was homogenized in 2 ml of cold saline. On the same day, the recipients (two days old) were fed an amount of homogenate equal to one-third of a donor brain. Starting at four days of age they were given one 2-min trial per day, with both tunnels open. After each trial, if the chick had not responded within the allotted time it was picked up and placed with the stooges on the social side of the apparatus for 30 sec. At the end of 30 sec the chick was removed and replaced in its brooder. After each session, chick feed (Purina Startena) was available ad libitum for several hours if the chick was to be run the following day. In the event that it would not be run the next day, food was available until the night before the experimental session. Water was available at all times, regardless of the experimental schedule. After giving the birds four 2-min trials, one per day, over a six-day period, we decided that 2 min unguided trials would not result in acquisition of the detour response at a rate rapid enough to show a difference. This decision was based upon our previous data which showed the greatest transfer effect within the first four trials. At this point (after four single trials) the birds were switched over to a two-trial session, keeping each trial at 2 min. After two of these sessions (four trials) they were given two additional unguided sessions consisting of two 3-min trials. Considering their age (14–15 days old), and the fact that only half (4–6 out of 12 chicks per group) of the subjects were making the detour response after a total of 12 trials, we tried to salvage the subjects and use them in the next experiment as donors. In order to get as many chicks responding as we could, we put them back onto a 2 min trial and started to guide them to the tunnels. After a total of 19 trials the birds were killed and their brains divided in half, after removing the cerebellum, and homogenized in equal volumes of cold saline. The donor brains were divided into three groups: good responders to tunnel no. 1, good responders to tunnel no. 2 and poor responders. A chick was defined as a good responder if it made 14 or more responses out of 19 trials with no more than one response to the inappropriate tunnel. The poor responders were defined on a basis of six or fewer responses to either or both tunnels. These brains (one-half brain each) were fed to two-day-old recipients, and after a five-day delay the recipients were started on a 2 min, two trial/day guided schedule. Recipients of extract from poor responding donors were guided to alternate tunnels, the tunnel to which the first unsuccessful subject was guided being randomly chosen. Thereafter, alternation of tunnels to which birds were guided was continued throughout the remainder of the experiment. Since we were still looking for possible position-specificity, each recipient of brain from tunnel no. 1 donors (good responders) was guided to tunnel no. 2 after its first failure and was subsequently alternated. The same procedure was carried out with the recipients of tunnel no. 2 donors except they were guided on their first failure to tunnel no. 1. Although we initially guided the recipients of good responder brains to the inappropriate tunnel, we predicted an overriding effect if a position-specific transfer ensued.

Results. The results of the second experiment in which a two-tunnel detour task was used also supports the interpretation that a transfer effect is possible. Aside from the ordinate scale being slightly different (due to 2 min trials instead of 4 min trials) the shape of the curves is almost identical to the original one-tunnel detour performance curves (fig. 9.7). In addition, the effect again

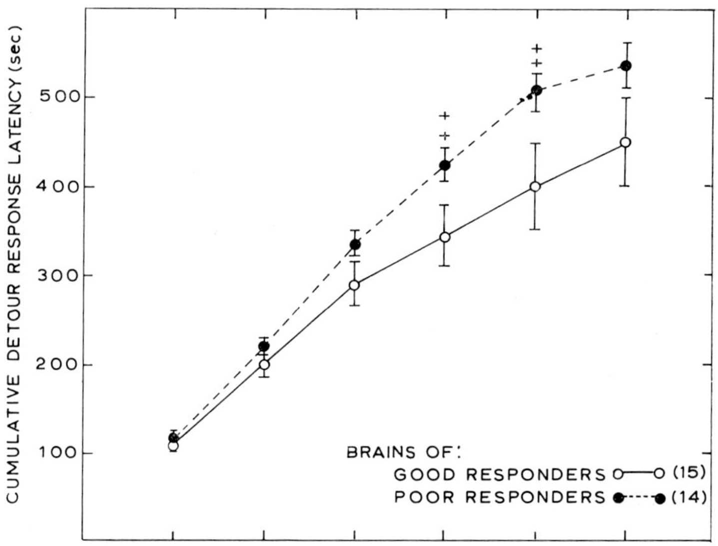

TRIAL	1	2	3	4	5	6
AGE (DAYS POSTHATCH)	7	7	8	8	9	9
RESPONSES 2 min						
GOOD RESPONDERS	3	6	9	10	11	12
POOR RESPONDERS	1	3	2	7	8	12

Fig. 9.7. Detour response latencies of groups receiving brains of good or poor responding donors. Good-responder recipients were combined, regardless of donor tunnel preference for comparison with poor-responder recipients. See legend to fig. 9.4 for explanation of symbols.

appears early in the series of trials and gradually diminishes with repeated trials. Although fig. 9.7 covers only the first six trials, given over a three-day period, the experiment was extended to cover five two-trial sessions. In this manner it was possible by the tenth trial to get each bird to make the detour response on its own at least once, except for one bird in the no. 1 tunnel recipient group. By looking at the choice of tunnel on the first response there appears to be a hint of position-specificity but only with the group that received brains from good responders to the mirror tunnel no. 1. Table 9.1 shows the distribution of tunnel choice for the three groups. Of the seven

TABLE 9.1

Choice of tunnel on first detour response.[a]

Recipients of brains from	Responsed to tunnel	
	no. 1	no. 2
A. Good responders to no. 1 tunnel [b]	6	1
B. Good responders to no. 2 tunnel	3	4
C. Poor responders	6	8

[a] It has been our experience that, once having responded, a chick will continue to respond to the same tunnel. By examining the first choice only, the problem of weighting values is negated.

[b] All values represent entire groups except for those in group A. One out of eight birds did not make the detour by ten trials. The probability that 6/7 will respond to one or the other tunnel, assuming a 50:50 choice if left to chance is < 0.0547 (binomial distribution for proportions).

subjects who did respond in the no. 1 tunnel recipient group, six chose the appropriate tunnel. That is to say, the probability that 6/7 responders will choose one tunnel over the other when the theoretical random choice ratio is 3.5/7 (50:50) is $p < 0.0547$. The other groups showed choice ratios near unity and did not approach statistical significance as did the no. 1 tunnel group.

Discussion. Our expectation of a position-specific effect was to no avail. The additional cues and increased size of the apparatus resulted in behavior which seemed to be incompatible with the detour response. If a position effect were to be realized, we could not guide them to a tunnel after their allotted time. After several sessions it became apparent that acquisition of the detour response without guidance would take considerably longer, especially if 2 min trials rather than 4 min trials were used. The use of shorter trials was necessitated by the expanded experimental design using four groups of 12 subjects each as was described in the first experiment with this apparatus. Whether or not this could account for our inability to show a difference in acquisition rates between experimental and control groups, regardless of specificity, is open to question. Results of other experiments, using the larger (two-tunnel) box, suggested that if a difference existed, it would be greatest during the initial trials, as was the case with the one-tunnel box. This finding was again borne out by the second transfer experiment in which the double tunnel box was used. A significant difference at the fourth and fifth trial between groups that received brains from good or poor responding donors was found. It must be remembered that the fourth trial occurred on the second session and the fifth trial on the third session, with differences diminishing and disappearing thereafter, as all subjects learned to respond.

By this time the question of unequal group size must have occurred to the reader. When using the very young of any species in an experimental design one should expect a greater incidence of deaths for a number of reasons. Our design employed oral administration of homogenate which further increased the probability of a few deaths due to aspiration of material into the lungs, etc. We have tried to keep group sizes constant or nearly constant, but occasionally a death or two resulted in unbalanced group sizes. Besides group size, the pertinent question of statistical handling of information and data remains a topic for real consideration. We have chosen to use cumulative response latencies because we feel there is a great dependence between current performance and an organism's behavioral history. Our decision is based upon the philosophy of the operant analysis of behavior, in which cumulative records are routinely used to measure effects of independent variables upon the maintenance or acquisition of behavioral repertoires.

9.3. Conclusion

The data which we have presented represent a fraction of that collected over the past year. Most of the information that has not been presented has been alluded to in procedural descriptions and discussions. Much time has been spent in determining the nature and sensitivity of the experimental model. From such information we have concluded that conditions and experimental design are of paramount importance. Although transfer experiments with the two-tunnel apparatus did not show a definitive effect they nevertheless were suggestive. At no time did we encounter significantly shorter latencies or more subjects responding in the control (stooge) or equivalent groups.

The suggestion of a transfer effect with another animal class and species adds to the generality of the phenomenon which in turn should encourage others to take an open minded view.

References

AGRANOFF, B. W., R. E. DAVIS and J. J. BRINK, 1965, Proc. Natl. Acad. Sci. U.S.A. 54, 788.

BAKAY, L., 1955, The blood–brain barrier. C. C. Thomas, Springfield.

BERGER, B. D. and L. STEIN, 1969, Psychopharm. (Berl.) *14*, 271.

BROWN, H., 1966, Psychol. Rec. *16*, 173.

BYRNE, W. L., D. SAMUEL, E. L. BENNETT, M. R. ROSENZWEIG, E. WASSERMANN, A. R. WAGNER, F. GARDNER, R. GALAMBOS, B. D. BERGER, D. L. MARGULES, R. L. FENICHEL, L. STEIN, J. S. CORSON, H. E. ENESCO, S. L. CHOROVER, C. E. HOLT, P. H. SCHILLER, L. CHIAPETTA, M. E. JARVIK, R. C. LEAF, J. D. DUTCHER, Z. P. HOROWITZ and P. CARLSON, 1966, Science *153*, 658.

COOK, L., A. B. DAVIDSON, D. J. DAVIS, H. GREEN and E. J. FELLOWS, 1963, Science *141*, 268.

CORNING, W. C. and E. R. JOHN, 1961, Science *134*, 1363.

FLEXNER, L. B., J. B. FLEXNER, R. B. ROBERTS and G. DE LA HABA, 1964, Proc. Natl. Acad. Sci. U.S.A. *52*, 1165.

FLEXNER, L. B., J. B. FLEXNER and R. B. ROBERTS, 1966, Proc. Natl. Acad. Sci. U.S.A. *56*, 730.

GORDON, N. W., G. G. DEANIN, H. I. LEONHARDT and R. H. GWINN, 1966, Am. J. Psychiat. *122*, 1174.

GROSS, C. G. and F. M. CAREY, 1965, Science *150*, 1749.

HYDÉN, H., 1963, Activation of nuclear RNA of neurons and glia in learning. *In*: D. P. Kimble, ed., The anatomy of memory, Proceedings of the first conference on learning, remembering and forgetting, 1965, vol. I. Science and Behavior Books, Palo Alto, pp. 178–239.

HYDÉN, H. and E. EGYHÁZI, 1962, Proc. Natl. Acad. Sci. U.S.A. *48*, 1366.

HYDÉN, H. and P. LANGE, 1965, Proc. Natl. Acad. Sci. U.S.A. *53*, 946.

KIMBLE, G. A., 1961, Hilgard and Marquis' conditioning and learning, 2nd. ed. Appleton-Century-Crofts, New York, pp. 406–407.

LAJTHA, A., 1957, J. Neurochem. *1*, 216.

LUTTGES, M., T. JOHNSON, C. BUCK, J. HOLLAND and J. MCGAUGH, 1966, Science *151*, 834.

MCCONNELL, J. V., 1962, J. Neuropsychiat. *3*, Suppl. 1, 542.

MCCONNELL, J. V., 1966, Ann. Rev. Physiol. *28*, 107.

SCHOLES, N. W., 1965, J. Comp. Physiol. Psychol. *60*, 114.

SIEGEL, P. R., 1965, Psychopharm. (Berl.) *12*, 68.

SOLYOM, L., C. BEAULIEU and H. E. ENESCO, 1966, Psychon. Sci. *6*, 341.

SPARBER, S. B. and F. E. SHIDEMAN, 1969, Develop. Psychobiol. *2*, 56.

UNGAR, G. and C. OCEGUERA-NAVARRO, 1965, Nature *207*, 301.

WAELSCH, H., 1965, The turnover of components of the developing brain; the blood–brain barrier. *In*: H. Waelsch, ed., Biochemistry of the developing nervous system, Proceedings of the first international neurochemical symposium, 1955. Academic Press, New York, pp. 187–201.

WAGNER, A. R., J. B. CARDER and W. W. BEATY, 1966, Psychon, Sci. *4*, 33.

WHITEHOUSE, J., 1964, J. Comp. Physiol. Psychol. *57*, 13.

CHAPTER 10

Chemical transfer of learned information in goldfish

G. F. DOMAGK* and H. P. ZIPPEL

Physiologisch-Chemisches Institut der Georg August Universität Göttingen Göttingen, Germany

10.1. Introduction

From various review articles (Frisch 1925, 1942; Herter 1953; Brown 1957; Teichmann 1962) it is evident that the chemical sense as well as vision are fully developed in fish. Frisch and his coworkers have published a large series of papers on experiments concerning the discrimination power and learning ability of fish. Recent experiments from this laboratory, performed on the goldfish (*Carassius auratus*), have allowed a detailed insight into the training behavior (taste, odor and color stimuli) during the single steps in the formation of a conditioned reflex (Bieck and Zippel 1969; Zippel 1970). By registering several behavioral components the level reached in the shock-free training procedure can be determined rather accurately (Zippel 1970b).

In experiments with goldfish the influence of various drugs on the short term and long term memory has been studied (Agranoff 1967; Agranoff et al. 1968); in these experiments the training of the animals was performed in a shock avoidance procedure. In our own shock-free goldfish experiments the extinction of memory following a completed training has been studied in test periods of several months (Zippel et al. 1970).

A large number of experiments showing transfer of acquired information from trained donors into naive recipients has been published during recent years. Positive results have been obtained with invertebrates (McConnell 1962) as well as with vertebrates (Babich et al. 1965; Ungar and Irvin 1967; Rosenthal and Sparber 1968). Among the procedure that have given positive results were cannibalism (McConnell 1962), injections of whole homogenates of brain (Rosenblatt 1969), of purified brain RNA (Røigaard-Petersen et al. 1968) or peptides prepared from the brains of donors (Ungar et al. 1968).

* Present adress: Department of Biochemistry, University of Louvain, Louvain, Belgium.

10.2. Methods

The fish (*Carassius auratus*), 12 to 15 cm in length, were kept in groups of two in training tanks (size: 130 × 30 × 20 cm). With our experimental setup good training results could be obtained only if the fish after being brought to the laboratory had been kept for six weeks in a big tank and then for at least four weeks had been adapted to the test situation (feeding of Tubifex through the plastic funnels). Insufficiently adapted animals did not show the normal avoidance reactions. Groups of two animals were found to be trainable in much shorter periods than were single fish. A shock-free training procedure (fig. 10.1) was used with green light, acetic acid, or quinine as the conditioned discriminative stimulus; these stimuli normally are repellent to fish (Bieck and Zippel 1969; Zippel and Domagk 1971). One training or testing session of 5 to 7 min was given per day, during which green light, acetic acid (concentration of 0.03 to 0.07‰) or quinine (concentration of 0.02 to 0.05‰), respectively, was retained in the close neighbourhood of one of the funnels. Each funnel had a constant influx of fresh tap water of about 15 °C (50 ml/min). In training, feeding with Tubifex worms through the small holes in the funnels was the reinforcement (unconditioned stimulus) offered with the conditioned stimulus. In five to ten training sessions all fish were found to react positively to the conditioned stimulus, and the training was continued for one more week. Following this a differentiation training was given with red light or glucose (final concentration 0.1 to 0.5‰) as an attractive concurrent stimulus (Bieck and Zippel 1969; Zippel and Domagk 1971) being offered

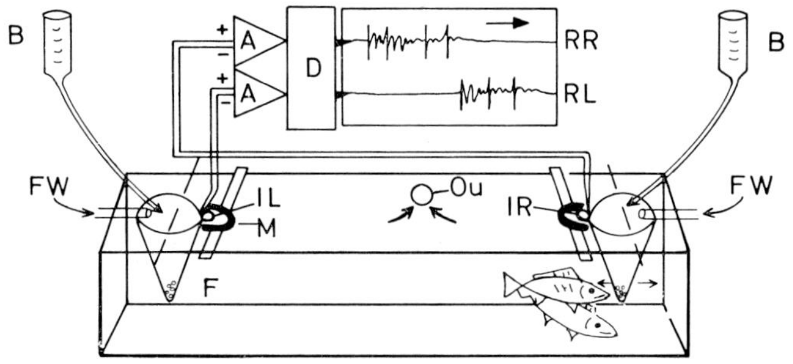

Fig. 10.1. Training apparatus used in goldfish experiments. Abbreviations used: A–amplifier; B–bottle with taste stimulus; D–directwriter; F–funnel; FW–fresh water supply pipe; IL, IR–induction coil at the left or right side of the tank; M–magnet; OU–outflow; RR, RL – registrations from the right and left funnel, respectively (Zippel and Domagk 1971a).

opposite to the conditioned stimulus. Biting (=food expectation) or swimming against the funnel was registered as either 'right' or 'left' funnel movement on a two-channel automatic recorder. Fish trained to a criterion of 80 to 90% correct were used as the donors. They were narcotized with Tricain Sandoz (MS 222) and their brains were removed as fast as possible. Brains pooled from four to ten animals were kept for at least one day in a freezer at −15 °C; a storage of several weeks did not destroy the transferable biological activity.

Extracts were prepared according to Ungar et al. (1968): the brains were homogenized in a Potter–Elvehjem apparatus with 10 to 20 vol of distilled water. The whole homogenate was dialyzed for 20 hrs against 20 vol of distilled water. The retentate showed only little biological activity and was

TABLE 10.1

Flow of subjects through an experiment (Zippel and Domagk 1971).

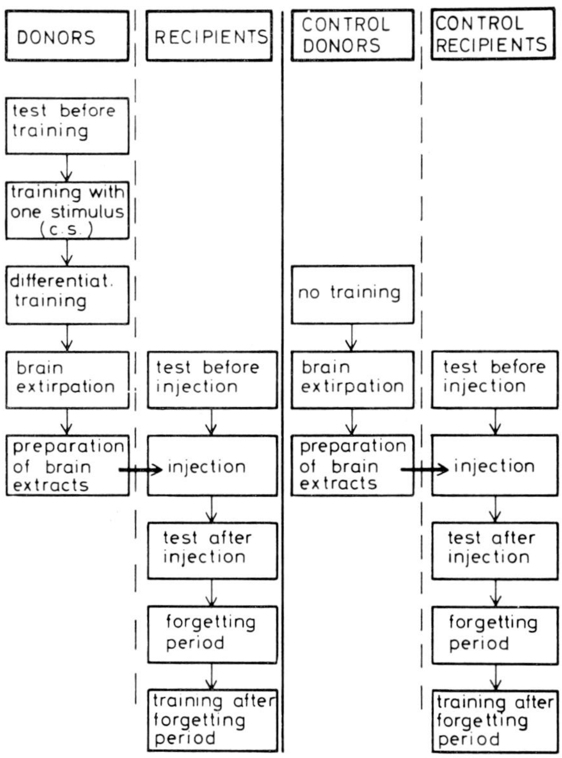

TABLE 10.2

Procedures used for training (1) and testing (2) of fish. The only differences occur under B3, where food (unconditioned stimulus) is offered during training, but not during the test sessions.

1: *Training of donors*	2: *Testing of recipients*
1A: *Tests before training*	2A: *Tests before injection*
1A1 and 2A1: Control registration without stimulus	
1A2 and 2A2: Test registration with conditioned stimulus offered at the spontaneously preferred side of the tank	
1A3 and 2A3: No food offered with conditioned stimulus	
1B: *Training*	2B: *Testing after injection*
1B1 and 2B1: Control registration without stimulus	
1B2 and 2B2: Test registration with conditioned stimulus offered at the spontaneously disliked side of the tank	
1B3: Food (unconditioned stimulus) with conditioned stimulus.	2B3: No food offered

discarded. The outer liquid was lyophilized and the very small residue dissolved in Ringer's solution. Control extracts were prepared by the same procedure using brains of untrained fish instead of trained donors.

The extract (usually 0.5 ml) was injected i.p. into naive, but pretested recipients, using a donor/recipient ratio of 1:1. After injection the recipients, also swimming in groups of two, were tested daily for their reaction towards the conditioned stimulus used in the training of the appropriate donors.

Table 10.1 gives a schematic presentation of the sequence of the various steps occurring in this study; table 10.2 demonstrates the close similarity between the training and the testing procedures.

10.3. Results

When the red light was offered to a pair of fish from the small side of a training tank, these animals would move to that side of the aquarium; green light, when offered to untrained fish, always was repellent.

All the fish used in this laboratory has been imported from Italy. When red and green lights were offered under comparable conditions to fish in the hands of an American investigator (Ungar 1969), curiously enough all these animals disliked the red light and showed a preference for green.

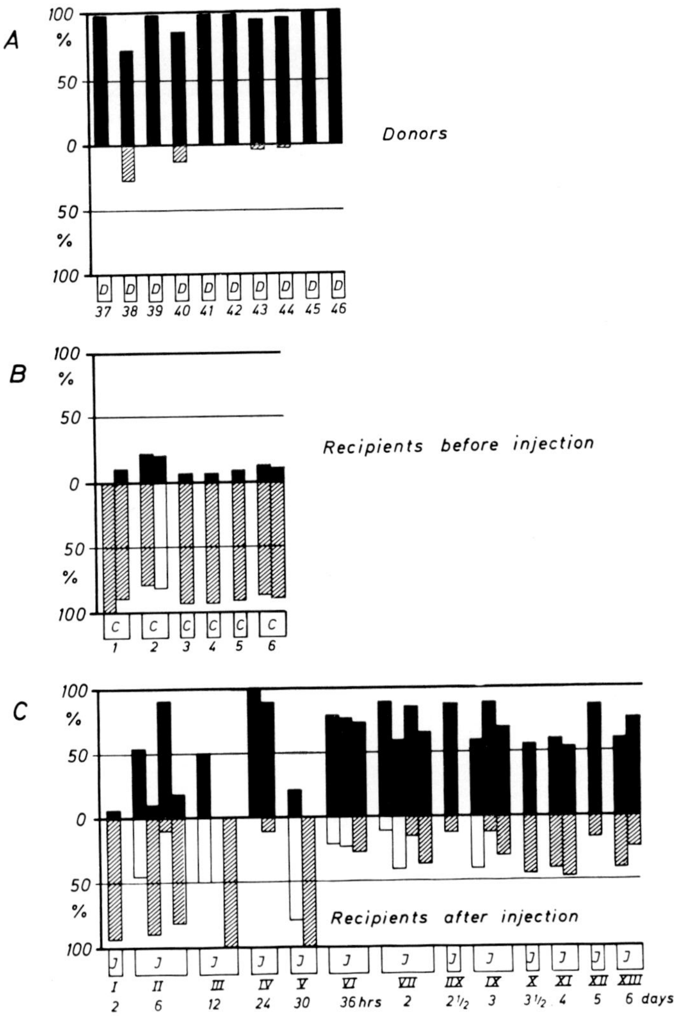

Fig. 10.2. Chemical transfer of color discrimination in goldfish. Symbols: □ preference to green light as compared to red light ▨ or to no light offered □. A: Behavior of a donor group (2 animals) exposed to green and red light after 37 to 46 training sessions: the fish have learned to prefer the green light. B: Behavior of a recipient group exposed to green and red light without previous training: preference of red light. C: Behavior of a recipient group from two hrs to six days after i.p. injection of a brain extract prepared from donor group A: there is a clear-cut change of preference.

The behavior of the recipients shown in fig. 10.2 is a typical example of the red preference observed in all naive animals. In the same figure, marked as donors, one sees how after about 40 training sessions, with food as the unconditioned stimulus offered after the conditioned stimulus, the fish show an almost complete green preference. Brain extracts prepared from such groups transfer the green preference to naive red-preferring recipients on i.p. injection. As is shown in the lower part of fig. 10.2, this green preference becomes apparent about 24 hrs after the injection and keeps for about one week. After complete forgetting the recipients can be trained for green light like untrained animals (see bottom of fig. 10.2). The i.p. injection of control extracts prepared by the same method from the brains of untrained animals does not alter the behavior of the recipients, as is shown in fig. 10.3. The specificity of these transfer effects will be discussed below.

As can be seen from figs. 10.4 and 10.5, before the injection all recipient groups showed the 'normal' behavior: acetic acid and quinine were repellent,

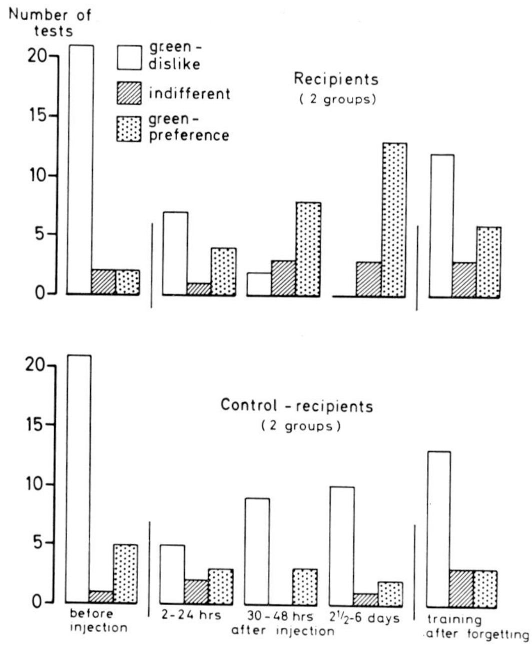

Fig. 10.3. Number of positive (green-preference), indifferent, or negative (green-dislike = red-preference) tests before and after injection of brain extract (Zippel and Domagk 1969).

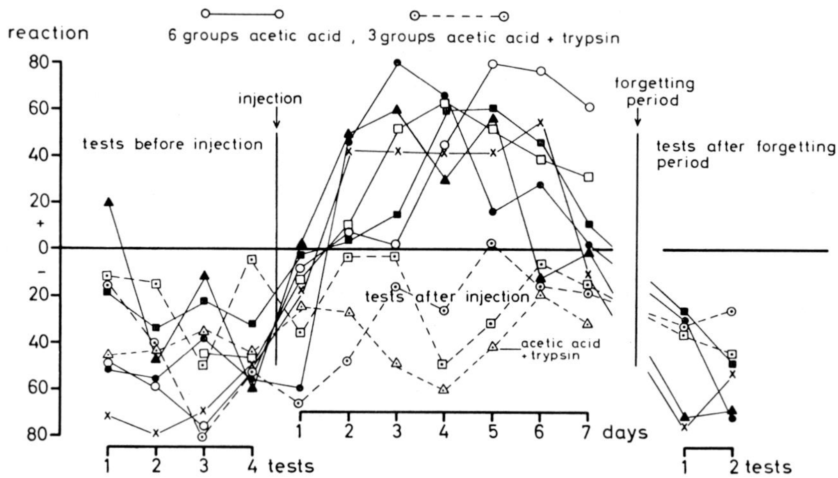

Fig. 10.4. Behavior of recipients after injection of brain extracts prepared from acetic acid trained donors. 'reaction' = biting and swimming movements in percent at the positive funnel (conditioned stimulus = acetic acid) compared to the control reactions at the same funnel without stimulus (see table 10.1). Positive values represent a preference for acetic acid, negative values show a dislike of this conditioned stimulus (Zippel and Domagk 1971a).

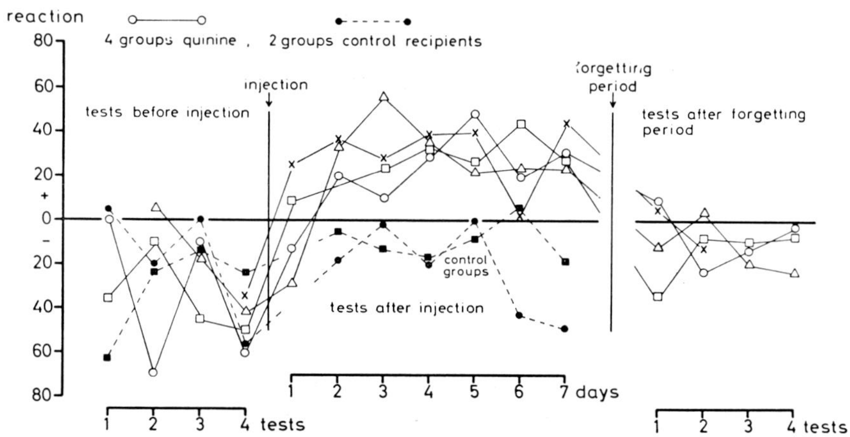

Fig. 10.5. Behavior of recipients injected with brain extracts from quinine trained fish (for details see legend to fig. 10.4; Zippel and Domagk 1971a).

whereas glucose solution had an attractive effect. Eight groups were injected with extracts prepared from acetic acid trained donors, four groups with extracts from 'quinine donors', and two groups served as controls. Up to 24 hrs after the injections almost all recipients still showed the strong dislike for the offered taste qualities, but a remarkable increase in general activity (number of bites, swimming movements) was observed after the 12th hr; this activity seems to be the first sign of a positive change in behavior and lasts for about one week. On the second day the originally repellent substances become liked by some of the fish, and then for a period of two to six days after the injections the behavior of the recipients is comparable to that of the donor groups during the training period. The positive behavior after injection can be observed for about one week; for the following three days there is only a slight preference for the taste offered as conditioned stimulus to the donors, and after two to three weeks after the injections the recipients react like normal controls with a strong dislike for the conditioned stimulus.

As had been shown in figs. 10.4 and 10.5 the first positive results in the transfer of taste training can be observed about 24 to 48 hrs after the injection. Before that time or after the extinction of memory (seven to ten days after injection) the injected recipients reacted to competing taste stimuli like naive animals: acetic acid and quinine were disliked, glucose acted attractive.

In fig. 10.6D to K, the behavior of a group of recipients injected with brain extract from acetic acid trained donors is shown. As is seen from line D the fish spontaneously preferred the left side of the tank. When the originally disliked acetic acid was offered now on the right side of the aquarium the fish moved over to this side as is shown under line E. When a spontaneously attracting glucose solution (fig. 10.6, line C) was offered at the left side of the tank together with the acetic acid offered at the right side, the behavior of the fish becomes uncertain but there is no longer the definite preference for glucose seen under 10.6C. In fig. 10.6, line G, a new control was performed, the fish again preferred spontaneously the left side of the tank. If now quinine, another substance acting repellent before the injection, is offered at the left side of the tank (line J) no generalization of change in preference is observed: the quinine is strongly disliked. In the registration shown under line K, the acetic acid was offered at the right side with quinine at the left side of the tank: all the animals move again towards the acetic acid.

Fig. 10.7 shows similar experiments performed on recipients of extracts from quinine adapted donors. Now the recipients are attracted by the

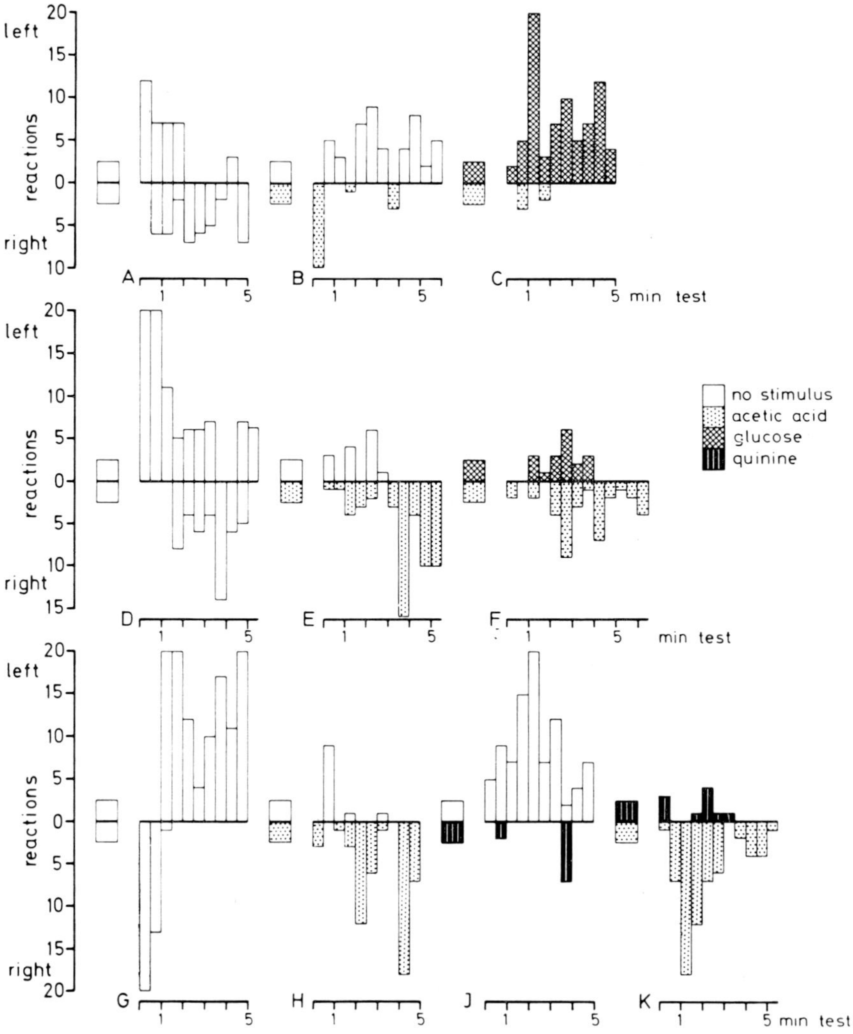

Fig. 10.6. Behavior of a recipient group exposed to acetic acid (conditioned stimulus), quinine and glucose (concurrent stimuli) before (A–C) and after (D–K) injection of a brain extract prepared from acetic acid trained donors (Zippel and Domagk 1971b).

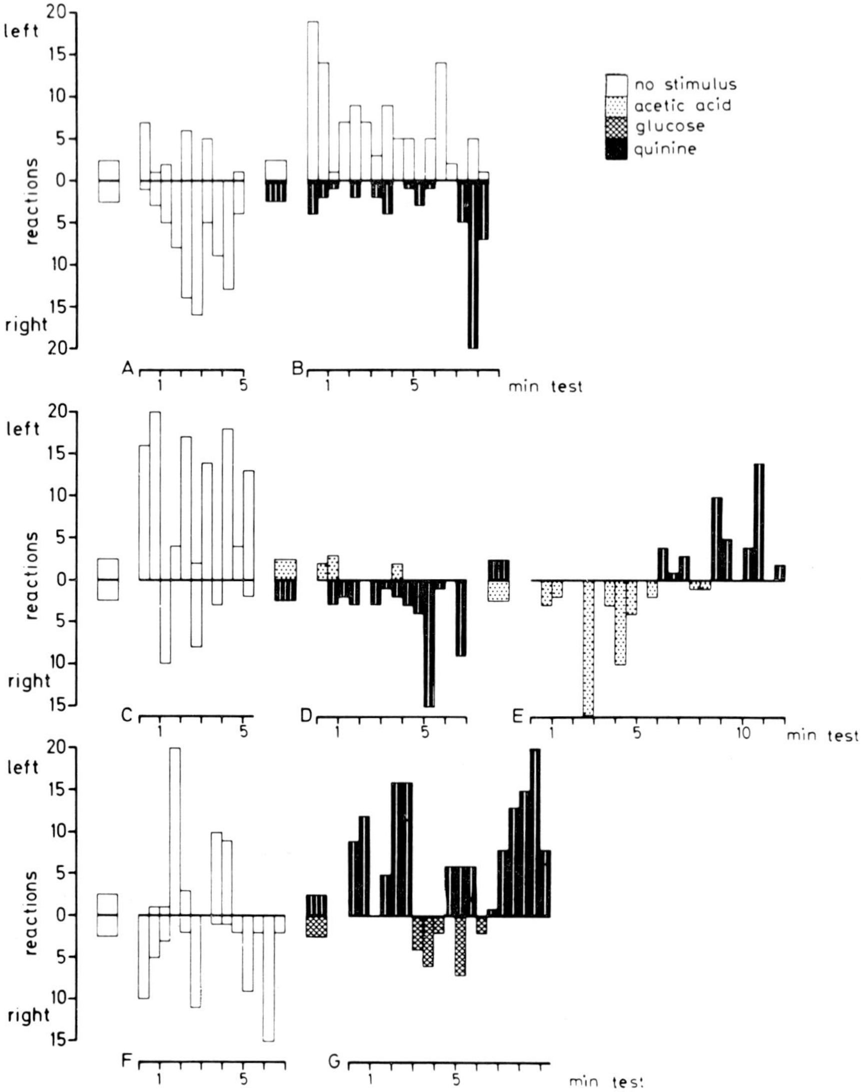

Fig. 10.7. Behavior of a recipient group exposed to quinine (conditioned stimulus), acetic acid and glucose (concurrent stimuli) before (A–B) and after (C–G) injection of a brain extract prepared from quinine trained donors (Zippel and Domagk 1971b).

spontaneously disliked quinine, which seems to be even more attractive than glucose; on the other hand acetic acid is repellent as before.

Fig. 10.8 shows the behavior of a group of recipients from acetic acid trained donors which five days after the injection were offered simultaneously taste and color vision stimuli. The control registration in line A without offering of any stimuli shows a slight preference to the right side of the

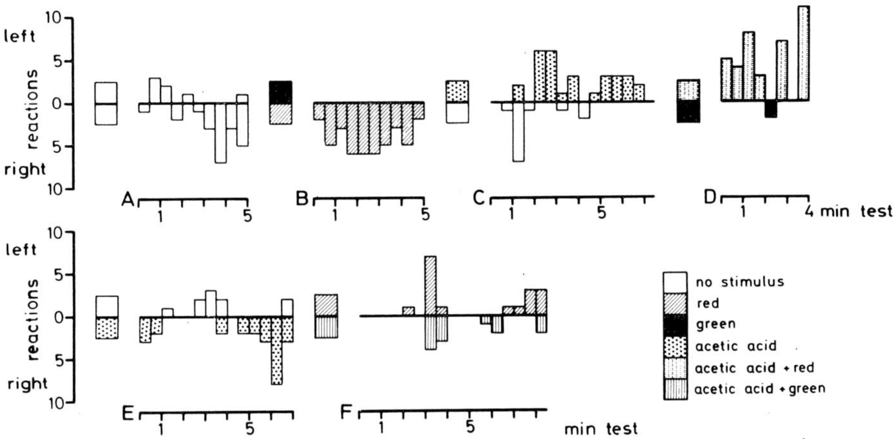

Fig. 10.8. Behavior of a recipient group exposed to acetic acid (conditioned stimulus) and concurrent taste and color stimuli after injection of a brain extract prepared from acetic acid trained donors (Zippel and Domagk 1971b).

aquarium. In line B red light was offered at the preferred and green light at the disliked side. A typical behavior of naive animals was observed: the fish approached exclusively the red light. The acetic acid offered at one side of the tank was found to be very attractive (line C). Red light offered simultaneously with acetic acid on one side with green light on the contralateral front increased the positive responses (line D). Subsequent an offering of acetic acid alone on the contralateral side gave a positive response after a few minutes of orientation. When, as shown in line F, acetic acid was offered together with green light with the attractive red light on the contralateral side the behavior of the fish became unpredictable.

A summary of the results of all experiments allows the following conclusion: recipients injected with a brain extract from acetic acid trained donors recognize the acetic acid as a positive stimulus between the second and tenth days. Positive responses are also observed with a similar offering of quinine (seven experiments: seven positive results) and glucose (18 experiments: fifteen positive, three negative results). Exposition of the donors to color stimuli alone gives control reactions: red light is attractive, green light acts repellent. The positive responses towards acetic acid are increased by a concomitant offering of red light whereas green light diminishes the positive reactions.

Recipients of extracts from quinine adapted donors react positively to quinine also under the simultaneous offering of acetic acid (nine experiments: nine positive results); in this series the positive results, however, were sometimes disturbed by the simultaneous offering of glucose (twelve experiments: seven positive, five negative results).

Conclusion: after injection of a brain extract from taste-adapted donors the recipients show a positive reaction not only towards the conditioning stimulus but they are able of distinguishing this from other tast stimuli. The injected information included not only the general signal 'taste = positive' but also gave a differentiation to the taste quality, e.g. 'acetic acid = positive'.

In order to get some informations on the chemical nature of the transfer active substance we have treated an aliquot of three biologically fully active extracts with trypsin. As can be seen from fig. 10.4 the transfer activity was abolished completely. From this fact and from the low molecular nature detectable in the dialysis step we assume, in agreement with Ungar (Ungar et al. 1968) that the active compound is a low molecular peptide. Fig. 10.9 shows the behavior of two groups towards acetic acid and glucose. The brains of four animals were pooled and purified as described above. Before the injection the material was divided into two parts: one half was injected i.p. into recipients as usual (fig. 10.9C to E), the other half was incubated at 37 °C with 20 μl of 0.1% trypsin and then injected into another group of recipients (fig. 10.9A and B). Five days after the injection the recipients of the trypsin-treated extract showed a spontaneous behavior like before the injection (fig. 10.9B): acetic acid is disliked, glucose is preferred. The recipients of the untreated aliquot of the extract react positively for acetic acid alone as well as for acetic acid offered simultaneously with glucose.

All the data mentioned above have undergone statistical treatment by the Wilcoxon test, the results of which are presented in tables 10.3 and 10.4

TABLE 10.3

Statistical analysis of data obtained from 15 fish groups (Wilcoxon test).

Conditioned stimulus	Tests before injection	Tests after injection			After forgetting	
		0–24 hrs	24–48 hrs	2–6 days	tests	training
Acetic acid (6 groups)	$p < 0.01$ dislike $n = 40$	$p > 0.075$ $n = 9$	$p < 0.01$ preference $n = 15$	$p < 0.01$ $n = 38$	$p < 0.01$ dislike $n = 16$	$p < 0.02$ preference $n = 6$
Quinine (4 groups)	$p < 0.01$ dislike $n = 17$	$p > 0.075$ $n = 4$	$p < 0.05$ preference $n = 8$	$p < 0.01$ $n = 20$		$p > 0.075$ $n = 10$
Controls (2 groups) (quinine)	$p < 0.01$ dislike $n = 14$		$p > 0.075$ $n = 4$	$p > 0.075$ $n = 10$		
Acetic acid (trypsin treated) (3 groups)	$p < 0.01$ dislike $n = 10$	$p < 0.05$ dislike $n = 5$	$p < 0.05$ dislike $n = 5$	$p < 0.01$ dislike $n = 14$	$p < 0.02$ dislike $n = 8$	$p < 0.025$ preference $n = 6$

TABLE 10.4

Statistical analysis of experiments performed with 4 fish groups (Wilcoxon test).

Before injection		After injection 0–24 hrs	After injection 24–48 hrs	After injection 2–6 days	Training after forgetting
$p < 0.02$ preference for red light, $n = 7$	Recipient group	$p > 0.075$ $n = 5$	$p < 0.05$ preference for green light $n = 5$ (both recipient groups)	$p < 0.02$ preference for green light, $n = 6$	$p > 0.075$ $n = 10$
$p < 0.01$ $n = 8$	Recipient group	$p > 0.075$ $n = 6$		$p < 0.05$ preference for green light $n = 6$	$p > 0.075$ $n = 10$
$p < 0.01$ preference for red light, $n = 9$	Control group	$p > 0.075$ $n = 5$	$p > 0.075$ $n = 5$ (both control groups)	$p > 0.075$ $n = 6$	$p > 0.075$ $n = 10$
$p < 0.01$ $n = 8$	Control group	$p > 0.075$ $n = 6$		$p < 0.05$ preference for red light $n = 5$	$p > 0.075$ $n = 10$

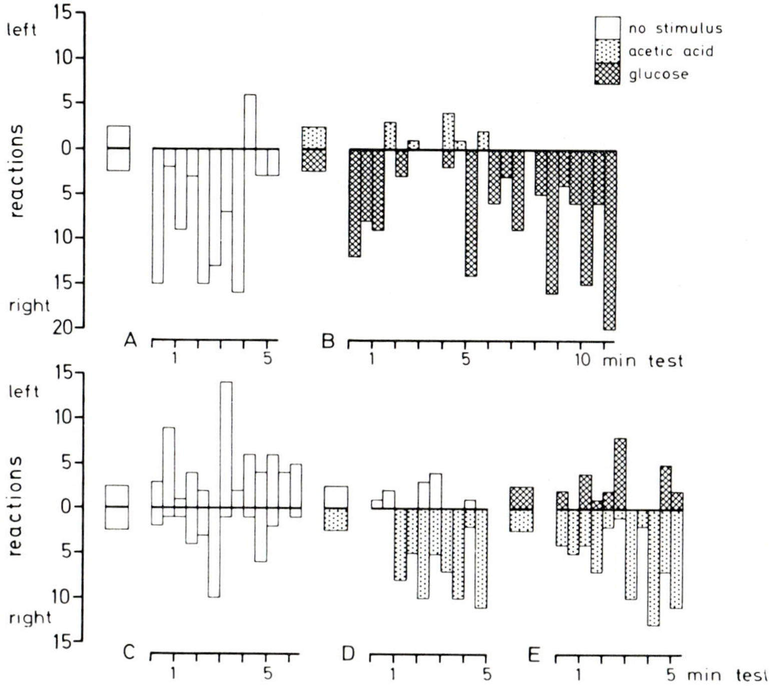

Fig. 10.9. Behavior of two recipient groups exposed to acetic acid (conditioned stimulus) and glucose after injection of a brain extract prepared from acetic acid trained donors. A, B: brain extract treated with trypsin; C, D, E: brain extract without trypsin treatment (Zippel and Domagk 1971b).

10.4. Outlook

Just a few years ago large groups of serious investigators were convinced that a chemical transfer of acquired information was not possible. Recently the number of research papers presenting positive results in such efforts has been increasing rapidly. Mostly rodents were used in these studies, and the transfer phenomenon was shown not to be species-specific, since cross transfers from one species into another have been successful in various instances. The results presented here give clear evidence for the possibilities of transfer of acquired information in fish. We have started experiments to show a 'double transfer' (the donors being trained to an optical as well as to a taste stimulus), and preliminary studies performed in our laboratory have given evidence that also certain inborn behavioral patterns can be transferred.

References

AGRANOFF, B. W., 1967, Sci. Am. *216*, 115.

AGRANOFF, B. W. and R. E. DAVIS, 1968, The use of fishes in studies on memory formation. *In*: D. Ingle, ed., The central nervous system and fish behavior. Univ. of Chicago Press, pp. 193–201.

BABICH, F. R., A. L. JACOBSON, S. BUBASH and A. JACOBSON, 1965, Science *149*, 656.

BIECK, B. and H. P. ZIPPEL, 1969, Pflügers Arch. 307, *2*, 132.

BROWN, M. E., 1957, The physiology of fishes, vol. II. Behavior. Acad. Press Inc., New York.

MCCONNELL, J. V., 1962, J. Neuropsychiat. *3* (Suppl. 1) 42.

MCCORNACK, R. L., 1965, Am. Stat. Ass. J. *60*, 864.

FRISCH, K. V., 1925, Z. vergl. Physiol. *2*, 393.

FRISCH, K. V., 1942, Umschau Wiss. Technik *17*, 260.

HERTER, K., 1953, Die Fischdressuren und ihre sinnesphysiologischen Grundlagen. Akademie Verlag, Berlin.

RØIGAARD-PETERSEN, H. H., T. NISSEN and E. J. FJERDINGSTAD, 1968, Scand. J. Psychol. *9*, 1.

ROSENBLATT, F., 1969, Proc. Natl. Acad. Sci. *64*, 2, 661.

ROSENTHAL, E. and S, B. SPARBER, 1968, The Pharmacologist *10*, 168.

TEICHMANN, H., 1962, Ergebn. Biol. *25*, 177.

UNGAR, G., 1969, personal communication.

UNGAR, G., L. GALVAN and R. H. CLARK, 1968, Nature *217*, 1259.

UNGAR, G. and L. N. IRWIN, 1967, Nature 214, 453.

ZIPPEL. H. P., 1970, Z. vergl. Physiol. *69*, 54.

ZIPPEL, H. P., B. BIECK, E. BENNEKEMPER-KNOOP and G. F. DOMAGK, 1970, Die unterschiedliche Beständigkeit der biologischen Information nach Dressur oder Injektion von Gehirnextrakten, in preparation.

ZIPPEL, H. P. and G. F. DOMAGK, 1969, Experientia *25*, 938.

ZIPPEL, H. P. and G. F. DOMAGK, 1971a, Pflügers Arch. *323*, 258.

ZIPPEL, H. P. and G. F. DOMAGK, 1971b, Pflügers Arch. *323*, 265.

ZIPPEL, H. P. and G. F. DOMAGK, 1971, J. biol. psychol., in press.

CHAPTER 11

The goldfish as an experimental subject in chemical transfer

EJNAR J. FJERDINGSTAD

Institute of General Zoology, University of Copenhagen
Copenhagen, Denmark

11.1. Introduction

Until recently investigations of the transfer phenomenon have been limited to two groups of organisms that are at extreme ends of the evolutionary scale, the planarian flatworms and the rodent mammals. The results of Rosenthal and Sparber (1968; see also chapter 9 in this volume) show that the effect may be obtained in birds, and some preliminary results by Shapiro (1969) indicate that this may be true of reptiles too. Therefore it seemed worthwhile to investigate whether a transfer effect could also be found in fish. Although such results had not been published*, a great deal of work had been done on learning in fish (see Ingle 1968), including the effects of metabolic inhibitors such as puromycin, acetoxycycloheximide and actinomycin on the learning process. These agents selectively inhibit the synthesis of certain classes of macromolecules and they concomitantly inhibit consolidation (see Agranoff and Davis 1968), but they do not interfere with the acquisition of the task. In the studies of Agranoff and his co-workers a water-filled shuttle box was used, with a light as conditioned stimulus and a low voltage dc shock as unconditioned stimulus. The fish were given one session, consisting of 20 trials, and after injection with the inhibitor were given a second session after 72 hrs. When compared with uninjected control fish, such fish showed a decreased retention of the avoidance training if the inhibitor was injected within an hour from the end of training. This agrees well with the results in mammals described by other authors (Flexner et al. 1967; Barondes and

* However, independently of the author, Drs. G. F. Domagk and H. P. Zippel of the University of Göttingen were at the same time obtaining transfer results in goldfish using a different behavioral and biochemical approach. See Zippel and Domagk (1969) and chapter 10 in this volume.

Cohen 1968) and indicates that memory in fish may be dependent on macromolecular synthesis. Further support for such a hypothesis comes from studies by Shashoua (1968) in which the learning of a fairly complex motor task was found to cause large changes in goldfish neuronal RNA, the ratio of uridine:cytidine increasing on the average more than 90% over that of controls. Puromycin injected before training not only blocked retention of the task but also the appearance of base ratio changes, although it did not interfere with learning the task. Rapoport and Daginawala (1968) found changes in neuronal nuclear RNA of catfish when perfusing the nasal openings with olfactory stimulants such as morpholine, shrimp extract and redfish extract. These changes were specific to the type of stimulant used, and were not found when applying strong irritants such as menthol. Several of the experiments were run with a split brain preparation in which one half of the brain served as an unstimulated control.

11.2. Methods

On deciding to test for the presence of transfer phenomena in fish (in making this decision the author was further motivated by the development of a severe allergy to rodent hair) the approach of Agranoff and Davis (1968) was chosen for training the fish and injecting the extracts. Minor deviations from their procedure were made which led to significant changes in the rate of learning, and therefore the procedure will be described in detail.

The fish used in these studies were 3 in common goldfish (*Carassius auratus*) obtained from Ozark Fisheries, Stoutland, Mo., the same source used by Agranoff. They were housed at 20 °C, three fish to a $2\frac{1}{2}$ gallon tank without aeration, the water being changed twice a week. One drop of Wardley's Superchlor (active ingredient: thiosulfate) was added per gallon to counteract chlorine in the city water. The fish were fed once a day with as much Wardley's Conditioner Goldfish Food as they would consume within a few minutes.

Fig. 11.1 shows the shuttle box used in the experiments. It is 12.5 cm wide, 27.5 cm long and 12.5 cm deep, being 4 cm narrower and shorter at the bottom. During training it was filled with 4 l of water to which was added the same chlorine reducing agent used in the tanks. A wall of black plexiglass divided the box into identical halves, leaving a space of 2.8 cm at the bottom through which the fish could shuttle between compartments, thereby activating a pair of infrared photocells (Lehigh Valley No. 221-10). The lateral walls of the compartments were completely covered by the electrodes made of 'monel' metal wire mesh. Through these electrodes an intermittent 22.5 mA dc current could be applied as unconditioned stimulus. Stimulus lights (2 W)

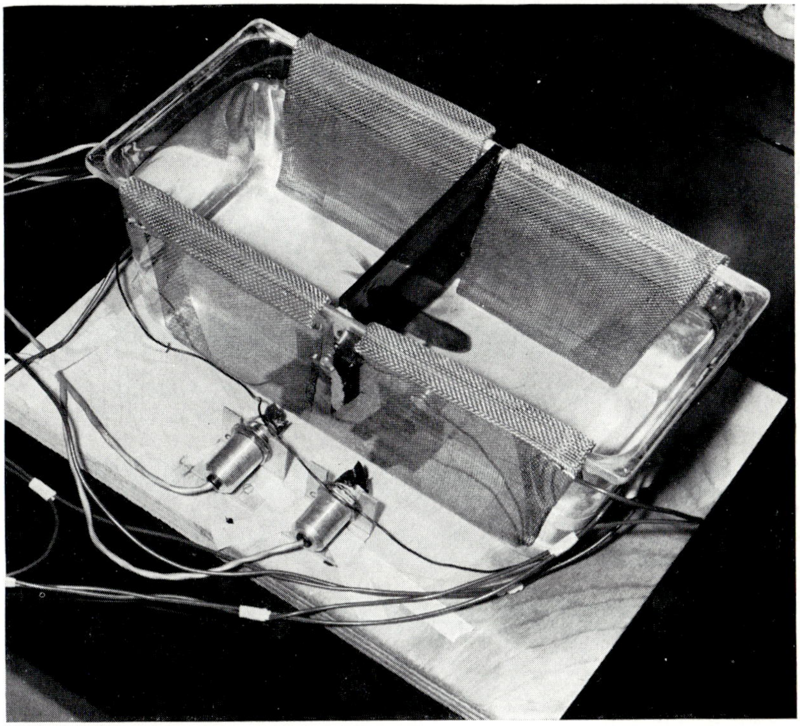

Fig. 11.1. The shuttle box used in the experiments. A fish is about to cross under the barrier separating the compartments.

to serve as conditioned stimulus were mounted on the end wall of each compartment. The setup was fully automated and during training was kept under a light-proof cover. Varying schedules of donor training and recipient testing were used in different experiments and will be described under each individual experiment.

Three to six hrs after the donor fish had finished their last training session they were killed by cutting the neck with scissors and the brain was removed and immediately frozen on dry ice. Pooled brains were extracted to yield the 'RNA extract II' described by Røigaard-Petersen et al. (1968). This extract contains about 10% protein and similar amounts of DNA as contamination. About 25 μg RNA, as determined by UV absorption, was obtained per brain (one brain = 50 mg wet weight). Similar (control) extracts were prepared from naive fish.

In all experiments the extracts were injected intracranially above the optic lobe according to the procedure described by Agranoff and Davis (1968). Twenty-five μg (experiments I, II and IV) or 30 μg (experiment III) dissolved in 10 ml of distilled water was injected with a Hamilton microliter syringe. Recipients of trained extract (experimentals) and of naive extract (controls) always received the same amount of RNA within any single experiment.

Testing of recipient fish always started 24 hrs after injection and was repeated every 24 hrs for several days. In all experiments testing was unreinforced, i.e. the fish were simply presented with intermittent periods of the stimulus light in the occupied compartment, and the number of avoidance responses were recorded. Thus, the possibility that the observed effect could be due to enhancement of learning or increased sensitivity to the unconditioned stimulus was eliminated.

A total of four experiments was run. Minor changes in procedure were introduced in each successive experiment in order to optimize the conditions for transfer.

11.3. Results

In the first experiment donors were given 20 trials a day. Each trial consisted of a 20 sec presentation of the conditioned stimulus, followed by 20 sec with both conditioned stimulus and unconditioned stimulus on, after which there was a 20 sec interval. The fish could avoid shock by shuttling in the conditioned stimulus period or escape by shuttling in the unconditioned stimulus period. The setup, however, was not programmed to be response contingent. The light and the shock remained on in the start compartment for the full duration, irrespective of whether there was a response or not. After two days with a performance of more than 80% avoidance out of trials, or after 10 days of training, whichever came first, the fish were sacrificed. The majority of the fish reached criterion within 10 days.

Recipient groups of ten and nine goldfish respectively were injected intracranially with 25 μg RNA extract from the donors. Twenty-four hrs later they were given ten 20-sec exposures to the stimulus lights, one per min, and this was repeated every 24 hrs for four days. The performance was recorded as cumulative number of responses, and after the second session there was a significant difference between the experimentals and the control fish (fig. 11.2). It was noted (this pilot study was not run blindly) that the difference was even more striking when considering only the first half of each test, i.e. the first five presentations of the light, and on the following two days the fish were only given five trials instead of ten. Fig. 11.3 shows the result obtained when considering only the first five trials.

Experiment I thus seemed to indicate that quite striking transfer might be obtained in fish by using the present approach. The fact that the effect was greater when considering only the first half of the tests might be due to rapid extinction of the transferred response and subsequent spontaneous recovery before the next test.

Experiment II used the same method of donor training except that all donor fish were given 10 days of training rather than being sacrificed after

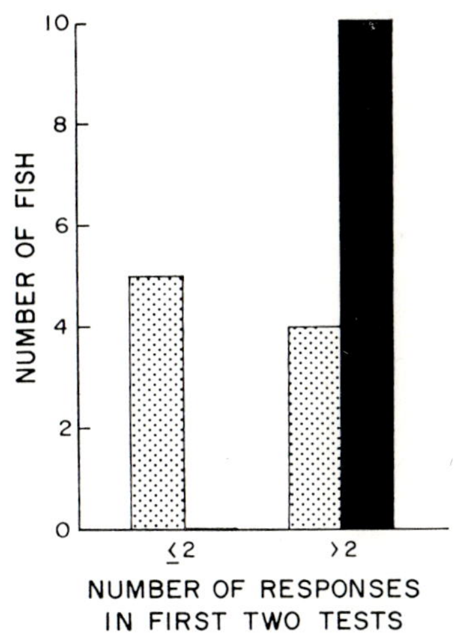

Fig. 11.2. Performance of the recipient groups in the first experiment. Black column: recipients of trained extract (n = 10, mean no. of responses = 8.4); stippled columns: recipients of control extract (n = 9, mean no. of responses = 6.1). The difference is significant according to the Fisher test, p = 0.025 (one-tailed).

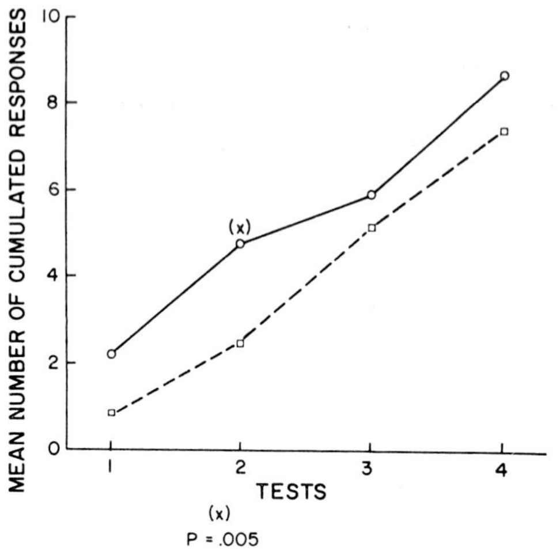

Fig. 11.3. Performance of recipient groups on first five trials of each session (see text). ○—○ recipients of trained extract (n = 10); □---□ recipients of control extract (n = 9). The Fisher test was used to evaluate the data.

reaching any predetermined criterion. Within the 10 days the mean performance of the donor group rose from about 20% avoidance on the first day to more than 80% on the last day (fig. 11.4).

Recipients of experiment II were injected with 25 μg RNA as in the preceding experiment. There were 12 fish in the control group, and 18 fish in the experimental group. Testing was done blindly, and again started after 24 hrs, but this time only five trials were given from the very first test. Again there was a difference between the groups in the predicted direction that reached significance on the fourth day of testing (fig. 11.5). Whether the delay in the appearance of the effect is due to the use of five trials instead of ten during the first two tests is not easy to say, but it remains interesting that the intergroup difference became significant after the same number of trials as in the first experiment.

In experiment III, donor fish were again given 10 trials a day for 10 days, but now the shock frequency was increased from one shock/sec to five shocks/sec during the unconditioned stimulus period. Though this obviously was more painful to the fish, no increase in rate of learning or final performance was apparent. Recipients were injected with 30 μg of RNA in both experimental and control groups, both of which consisted of nine fish. Twenty-four hrs before injection, the fish had been screened for preinjection level of responding by giving them twenty 10-sec unreinforced presentations of the light similar to the testing procedure. Fish that made more than four

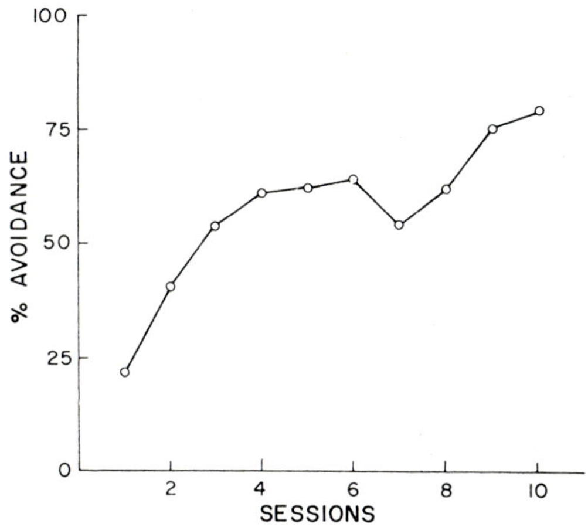

Fig. 11.4. Learning curve of 20 donor goldfish given 20 trials per day (there was an interval of 48 hrs between session six and seven). The duration of the conditioned stimulus period was 20 sec.

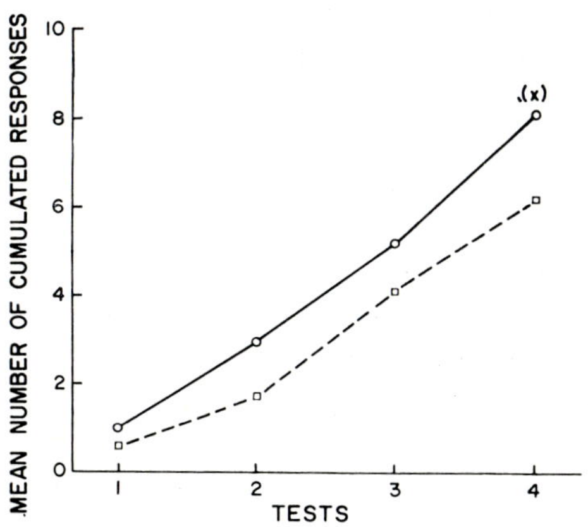

Fig. 11.5. Performance of recipient groups in the second experiment. ○—○ experimentals (n = 18); □---□ controls (n = 12). The larger number of recipients allowed the use of the χ^2 test (one-tailed, df = 1) for the evaluation of the data.

avoidance responses were rejected; this was about 25% of the total number of fish.

Testing in experiment III was blind as in the preceding experiment and the fish were given 10 trials a day. The level of responding was lower in this experiment, and the intergroup differences smaller. However, on the second test there was a significant difference in the number of responses in the predicted direction (fig. 11.6). It would thus seem that the screening procedure used was rather disadvantageous, most likely because it yields a relatively larger number of totally inactive fish in the groups.

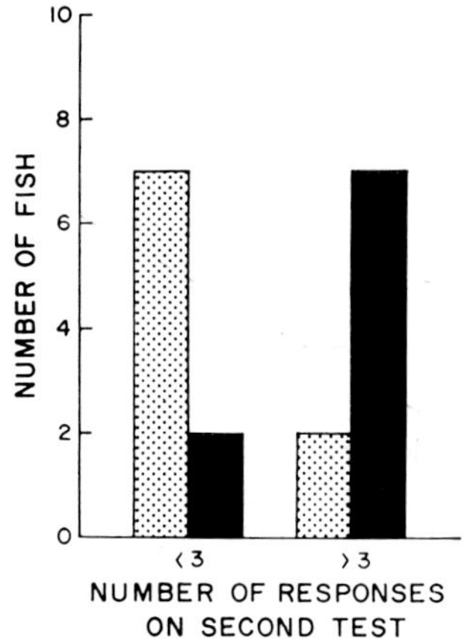

Fig. 11.6. Performance on the second test of recipients of the third experiment. Black columns: experimentals ($n = 9$, mean $= 3.89$); stippled columns: controls ($n = 9$, mean $=$ 2.78). The difference was significant according to the Fisher test, $p = 0.05$ (one-tailed).

In experiment IV, major changes were carried out in the donor training procedure. It would seem that the long (20 sec) conditioned stimulus period in the original Agranoff procedure would be somewhat disadvantageous in transfer experiments, since it is obvious that some of the fish actually learn that there is a long 'safe' period from the onset of the conditioned stimulus to the onset of the unconditioned stimulus. Thus, after an initial decrease of latency to respond, an actual increase may be seen in individual fish during donor training. For transfer it would seem better that all the fish learn to react quickly to the conditioned stimulus. Also for the test it might be advantageous to use a short conditioned stimulus period, since the chance of a random response in the controls within this period would thus be minimized. Accordingly, the 'conditioned stimulus only' period was reduced from 20 to 5 sec in donor training, i.e. $\frac{1}{12}$ of a cycle instead of $\frac{1}{3}$. At the same time the 'conditioned stimulus plus unconditioned stimulus' period was changed from 20 sec to 15 sec and the interval increased to 40 sec. Other conditions remained the same as in the preceding experiment. As may be seen from fig. 11.7,

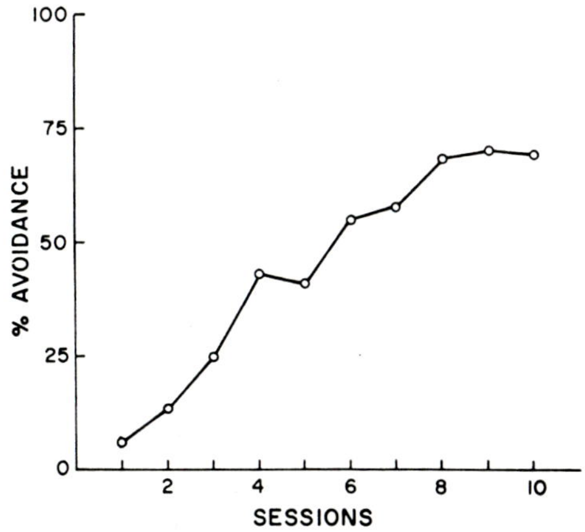

Fig. 11.7. Learning curve of 24 goldfish given 10 trials per day with a conditioned–unconditioned stimulus interval of only 5 sec (72 hrs elapsed between sessions four and five). On comparing with fig. 11.4 it is evident that the initial level of responding is more than three times lower, so that the relative increase in responding due to learning is much larger, i.e. more than ten times instead of less than four times.

the changes resulted in a much more striking learning curve. The performance on the first training session was now only 6% avoidance and this increased more than ten times during the total of ten sessions.

During testing in experiment IV the conditioned stimulus was presented for 5 sec only, once a minute. Perhaps for this reason the level of responding remained low, 10–15% avoidance as compared to 50% in the trained extract groups of earlier experiments. On the third of the individual sessions there was a highly significant intergroup difference in the predicted direction; however, on four out of five sessions there was a slightly higher number of responses in the control recipients, a tendency which was significant on the first day ($0.05 > p > 0.025$). This initial reversal, followed by a later effect in the predicted direction, is extremely similar to the results obtained by Røigaard-Petersen et al. (1968, see also chapter 4) in rats. It is interesting that it appeared in the goldfish experiment as the result of a relatively simple change of procedure.

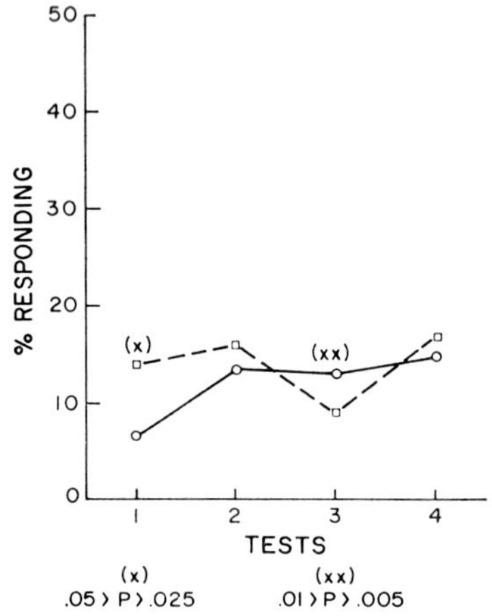

Fig. 11.8. Percent responding in recipients of the fourth shuttle box experiment, in which the duration of the conditioned stimulus presentation was also reduced from 20 sec to 5 sec. ○———○ experimentals (n = 17); □------□ controls (n = 10). An initial significant reversal is followed by a highly significant effect in the predicted direction. This fig. may be compared with the results obtained in several studies with rats (figs. 4.6 and 4.7). The results were evaluated by means of the χ^2 test (one-tailed, df = 1).

11.4. Discussion

The variations in procedure among experiments make it impossible to pool the data for statistical treatment. However, the fact that in all four experiments a significant to highly significant effect in the predicted direction was seen, while there was only one instance of (barely significant) reversal, shows that the overall reproducibility of chemical transfer in fish is certainly no less than in rats. Since the extracts used were originally developed for use in mammals, it would seem likely that the chemical nature of the active component is the same. However this, as well as the specificity of the effect in fish, must await further experimentation.

Apart from this, fish do have several unique advantages over rodents. For one thing, they have a human type color vision with three types of retinal cones so that it is possible to train them on color discrimination (see Yager 1968) and this might be advantageous for work on the specificity of transfer. Furthermore it is easy to control the temperature and oxygen tension of their environment. Some preliminary results in our laboratory indicate that fish can be trained in an anaerobic environment with oxygen as reinforcement. Manipulation of temperature in an analogous way has been used for training purposes by Rozin (1968), and because fish are poikilothermic this procedure may also be used to change the metabolic rate, which would appear potentially advantageous in studies on the formation and uptake of the active components. Finally, in times like the present it should not be overlooked that the cost of a fish experiment is going to be only one tenth of that of a rat study with a similar number of experimental subjects.

References

AGRANOFF, B. W. and R. E. DAVIS, 1968, The use of fishes in studies on memory formation. *In:* D. Ingle, ed., The central nervous system and fish behavior. University of Chicago Press, pp. 193–201.

BARONDES, S. H. and H. D. COHEN, 1968, Science *160*, 556.

FLEXNER, L. B., J. B. FLEXNER and R. ROBERTS, 1967, Science *155*, 1377.

INGLE, D., ed., 1968, The central nervous system and fish behavior. University of Chicago Press.

RAPOPORT, D. A. and H. F. DAGINAWALA, 1968, J. Neurochem. *15*, 991.

RØIGAARD-PETERSEN, H. H., T. NISSEN and E. J. FJERDINGSTAD, 1968, Scand. J. Psychol. *9*, 1.

ROSENTHAL, E. and S. B. SPARBER, 1968, The Pharmacologist *10*, 168.

ROZIN, P., 1968, The use of poikilothermy in the analysis of behavior. *In:* D. Ingle, ed., The central nervous system and fish behavior. University of Chicago Press, pp. 181–192.

SHAPIRO, H. C., 1969, personal communication.

SHASHOUA, V. E., 1968, Nature *217*, 238.

YAGER, D., 1969, Vision Res. *9*, 179.

ZIPPEL, H. P. and G. F. DOMAGK, 1969, Experientia *25*, 938.

Note added in proof

In further studies with the described behavioral technique, it has proved possible to train 12 or more fish at a time in a larger shuttle box, modified so that after onset the shock stays on in one compartment until the stimulus light for the next trial goes on in the opposite compartment. The 'mass trained extract' derived from these fish may then be evaluated either by testing individual recipients in the smaller shuttle box, or by massed testing in the large box. In the latter case, however, recording will have to be done by observation rather than automatically. The larger number of trained donors that may be obtained with this procedure should prove valuable in biochemical studies, e.g. the isolation of the active factor.

CHAPTER 12

Progress in the study of learning and chemical transfer in planarians

ALLAN L. JACOBSON

Department of Psychology, San Francisco State College, San Francisco, California

Some 15 years ago, Thompson and McConnell (1955) reported that they had successfully trained planarians in a simple Pavlovian-type learning situation. This claim, in itself, did not attract a great deal of attention at the time. In subsequent years, McConnell and his co-workers reported that this conditioning effect survived regeneration of a bisected worm (McConnell et al. 1959) and could be 'transferred' to planarians which cannibalized the trained worms (McConnell 1962). These more startling findings, which gained the planarian some modest fame in psychology, also led a number of psychologists and zoologists (as well as an occasional chemist or physicist) to examine McConnell's interpretations more closely. Critics of the work have argued, for example, that planarians are too 'lowly' an animal to be capable of learning, or that what appears to be classical conditioning is in fact a 'sensitization' effect. The nature of the chemical transfer effect has, of course, also been questioned, and the effect itself has been doubted by many.

As it often does, controversy on this problem has had the positive effect of stimulating investigators to do better and more conclusive research. Numerous experiments have been concerned with assessing the learning capacities of planarians: at one time or another, planarians have been trained to turn to the left or right side, or the lighted or dark side of a maze; to ignore a repeatedly presented stimulus; to break a small photobeam; to descend a wire; and, of course, to contract in response to certain cues. Each purported demonstration of learning has been attacked, and the literature on this topic is an interesting story in itself. Although I shall not in this paper review the general question of learning in planarians, I can refer the interested reader to any of several discussions on the subject (Jacobson 1963, 1965; Warren 1965; McConnell 1966a; Corning 1969).

I would like to focus this discussion on one particular form of learning, classical conditioning, since this form has been the one most used in studies of chemical transfer in planarians and most debated in discussions of planarian learning. Several recent studies have helped in clarifying and perhaps answering key questions in this area, and I will describe these studies in some detail, after reviewing the points at issue.

A brief pulse of current passed through water reliably evokes a gross bodily response in planarians. A burst of light evokes few such reponses. When the light repeatedly precedes the current, planarians respond more and more consistently to the light, prior to shock delivery. This was the training procedure devised by Thompson and McConnell (1955) and employed, in one variation or another, by many subsequent investigators. The response increase to light may represent classical conditioning. Another possibility is that through sheer stimulation, the worms become progressively more sensitive to the light. This would appear quite a different effect from classical conditioning, and has been labeled 'pseudo-conditioning' or 'sensitization'. There are subtle differences between these two terms which I have discussed elsewhere (Jacobson 1967). For the moment it is sufficient to consider them together as non-specific sensitizing effects which may be distinguishable from the specific association which classical conditioning conventionally implies. Thompson and McConnell recognized this problem, but the results of their control procedures led them to reject the sensitization interpretation of their data. Some critics, however, consider these control procedures inadequate (Jensen 1965). We need not analyze this particular disagreement, since further studies have provided clearer tests of the two interpretations.

The question then becomes: how can we distinguish empirically between classical and pseudo-conditioning effects? I.e., how can we determine whether the planarian is learning to respond to a signal, as it were, rather than becoming diffusely sensitized by cumulative stimulation? Two general methods have been employed. In one, amount of stimulation is equated for experimental and control groups, but the temporal relationships between the conditioned stimulus and unconditioned stimulus differ. For the experimental group, the conditioned stimulus (usually, with planarians, light) immediately precedes the unconditioned stimulus (usually shock) time after time. Thus, from our perspective, the light may be considered a reliable signal of impending shock. For the control group, an equal number of lights and shocks are presented over an equal time period, but the conditioned stimulus does not consistently precede the unconditioned stimulus. Several arrangements are possible in the latter case: the conditioned stimulus may consistently follow (rather than precede) the unconditioned stimulus; the conditioned stimulus and unconditioned stimulus onsets may be simultaneous; or the conditioned stimulus and unconditioned stimulus may be presented in a randomly interspersed sequence. If equal response increments were produced in experimental and control conditions, no claim of classical conditioning could be made. If, however, 'forward' pairing yielded appreciably greater response increments than control procedures, the pseudo-conditioning interpretation could be rejected.

The second method for comparing the two views is the use of a differential conditioning procedure. In this case, two different conditioned stimuli (usually light and vibration in the planarian work) are presented an equal number of times. One conditioned stimulus is always followed immediately by the unconditioned stimulus; the other conditioned stimulus never. Thus conditioned stimulus (+) (the stimulus paired with the unconditioned stimulus) 'signals' the unconditioned stimulus, whereas conditioned stimulus (−) does not. If, over the course of training, the planarians respond more to conditioned stimulus (+) than to conditioned stimulus (−), this represents a form of discrimination which is incompatible with the concept of non-specific sensitization.

These methods of recognizing 'true' classical conditioning are not peculiar to planarian conditioners; rather, they are typical of the views of psychologists expert in this area (see Gormezano 1966). Various investigators have employed one or another of these forms of controls in studies on planarians, typically with results which supported the classical conditioning interpretation. Forward conditioning (conditioned stimulus preceding unconditioned stimulus) has been found to produce greater responsiveness to the conditioned stimulus than does an interspersed order (Baxter and Kimmel 1963; Crawford et al. 1965; Kimmel and Yaremko 1966), a backward sequence (Vattano and Hullett 1964), or a simultaneous presentation (Vattano and Hullett 1964) of conditioned stimulus and unconditioned stimulus. Successful differential conditioning has been reported by Block and McConnell (1967), Fantl and Nevin (1965), and Griffard and Pierce (1964). Negative findings with differential conditioning were reported by Kimmel and Harrell (1964).

Sheldon Horowitz, Clifford Fried and I performed a series of experiments in which we approached this issue in several different ways. One subset of experiments varied the conditioned–unconditioned stimulus sequence, while the other subset attempted to establish differential conditioning. We used blind procedures wherever possible. In the first subset of experiments, different groups of planarians were subjected to forward conditioning, backward conditioning, simultaneous conditioning, or conditioned stimulus only. When the several training procedures were completed, we tested the responsiveness of the worms to the conditioned stimulus. In these tests, no shocks (unconditioned stimulus) were given, and the experimenter was unaware of the type of previous training received by each subject. When the conditioned stimulus was light, we obtained clear-cut differences between the experimental group on the one hand and the several control groups on the other. The results were more ambiguous when vibration was used as the conditioned stimulus: a low level of vibration yielded no consistent differences among groups, but the experimental group was clearly superior to controls when a somewhat higher level of vibration was used. The specific data in these experiments are presented in Jacobson et al. (1967). In general, the results provide strong support for a classical conditioning interpretation.

In the second subset of experiments, we conditioned worms differentially, using light and vibration as the conditioned stimuli. In half the cases, light was the conditioned stimulus (+) initially. In the second part of the training procedure, a reversal was instituted: the stimulus which was initially conditioned stimulus(+) became conditioned stimulus(−), and vice versa. Our results were quite striking: each of 12 planarians tested was able to differentiate appropriately between the two conditioned stimuli, during both the initial and the reversal stages of training. Because this study was not run blind, we performed an additional experiment to check for possible bias. Worms were given differential conditioning, half with light as conditioned stimulus(+) and half with vibration as conditioned stimulus(+), and were then tested with light and vibration (but no shock), in blind fashion. Each group proved to be more responsive to its respective conditioned stimulus (+), thus lending support to the findings obtained during training (again, see our 1967 paper for the data on these experiments). It is noteworthy that Block and McConnell (1967) also found differential responding in both initial and reversal stages of training.

Horowitz, Fried and I felt that the data of these two subsets of experiments constituted a strong case for the validity of classical conditioning in planarians.

The case is strengthened by the reports of other investigators whom I have cited. Problems remain unresolved in the area – the most pressing one is that many competent investigators are unsuccessful in their attempts to condition planarians. This is a serious problem, but one which can be considered separately from the present questions. The answers can only be found by a more detailed knowledge of specific factors influencing planarian conditioning (see McConnell 1967).

I would like to turn now to the question of chemical transfer of training in planarians. In 1962, a paper was published entitled 'Memory transfer through cannibalism in planarians' – surely one of psychology's more intriguing paper titles. In that study by James McConnell and his co-workers, planarians which had ingested conditioned planarians were found to be more responsive to the stimulus used in conditioning than members of a control group fed untrained planarians. The effect was impressive, but the interpretation difficult for several reasons. First, as has been discussed, there was much dispute over the validity of classical conditioning results in planarians. Secondly, and very much related to the first point, certain control procedures were not employed in the McConnell study which might have clarified the transfer phenomenon. Nonetheless, the experiment was a pioneering one and was the original report on chemical transfer.

Critics of the McConnell study pointed out that since control cannibals ate untrained worms, it was not clear whether the transfer effect observed was due to sensitization rather than conditioning. The same criticism applies to John's (1964) successful replication (which must still be valued as a confirmation, and a convincing one, of McConnell's findings). Hartry et al. (1964) found that sheer stimulation of 'victim' worms yielded as great a transfer effect as did conditioning of the victims. This study, however, had its own deficiencies (see Jacobson 1965). In any case, it was clear that better designs were necessary to resolve this question. One approach has been to utilize different learning situations. Thus Westerman (1963) reported successful cannibalistic transfer of habituation, and McConnell (1966b) has described stimulus-specific transfer effects with planarians trained to select the lighter or darker arm of a maze.

Fried, Horowitz and I (Jacobson et al. 1966) pursued the transfer effect with classical conditioning, using additional control procedures in an attempt to interpret the effect more adequately. We also used a different method of introducing material from trained worms into untrained worms. Rather than having the latter group eat the former, we prepared a chemical extract from the trained group and injected this extract into untrained worms. This procedure was introduced by McConnell and his co-workers (Zelman et al. 1963) and later used by Fried and Horowitz (1964). Both groups obtained transfer effects, although once again interpretation was difficult.

The experiment we performed, then, was designed to assess whether the transfer effect was dependent on conditioning of victim worms, or whether equated stimulation of victims would produce the same results. One group of planarians was classically conditioned with paired light and shock. A second group received the same number of lights and shocks, hence the same total stimulation, but the stimuli were unpaired. A third group received no stimulation. A chemical extraction procedure designed to secure ribonucleic acid (RNA) was then performed on each of the three groups. Next, the extract obtained was injected into three groups of untrained planarians. Finally, these recipient planarians were tested in the experimental situation some 24 hrs later. Two methodological points deserve mention here: (a) extraction, injection and testing were all conducted blind; (b) no shock was used during testing of recipients – their tendency to respond was assessed in the absence of explicit conditioning.

The results were quite clear: the experimental group responded to the light much more than either of the control groups, and the latter two groups did not differ. That is, worms which received extract from conditioned victims were clearly superior to worms which received extract from either stimulated or non-stimulated victims. Furthermore, during the original training of victims, conditioned worms were far more responsive to the light than were pseudo-conditioned worms. Thus the relationship of response levels obtained in original training of victims was 'recovered' in later testing of recipients. We performed several additional experiments (described in Jacobson et al. 1966) which corroborated and extended these basic findings.

These results speak for themselves: conditioning victim worms produces transfer effects markedly different from simply stimulating the victim worms. This points to a degree of specificity in the transfer phenomenon. Further information could be gathered about specificity if a differential conditioning procedure were employed with victim worms. This would be an elegant experiment, but technically a very difficult one.

References

BAXTER, R. and H. D. KIMMEL, 1963, Amer. J. Psychol. *76*, 665.

BLOCK, R. A. and J. V. MCCONNELL, 1967, Nature *215*, 1465.

CORNING, W. C. and D. RICCIO, 1970, The planarian controversy. *In:* W. L. Byrne, ed., Molecular approaches to learning and memory. Academic Press, New York.

CRAWFORD, F. T., F. J. KING and L. E. SIEBERT, 1965, Psychon. Sci. *2*, 49.

FANTL, S. and J. A. NEVIN, 1965, Worm Runner's Digest, *7*, 32.

FRIED, C. and S. D. HOROWITZ, 1964, Worm Runner's Digest *6*, 3.

GORMEZANO, I., 1966, Classical conditioning. *In:* J. J. Sidowski, ed., Experimental methods and instrumentation in psychology. McGraw-Hill, New York, p. 385.

GRIFFARD, C. D. and J. T. PIERCE, 1964, Science *144*, 1472.

HARTRY, A. L., W. D. MORTON and P. KEITH-LEE, 1964, Science *146*, 274.

JACOBSON, A. L., 1963, Psychol. Bull. *60*, 74.

JACOBSON, A. L., 1965, Anim. Behav. *13* (supplement), 76.

JACOBSON, A. L., 1967, Classical conditioning and the planarian. *In:* W. C. Corning and S. Ratner, eds., Chemistry of learning; invertebrate research. Plenum Press, New York, p. 195.

JACOBSON, A. L., C. FRIED and S. D. HOROWITZ, 1966, Nature *209*, 601.

JACOBSON, A. L., S. D. HOROWITZ and C. FRIED, 1967, J. Comp. Physiol. Psychol. *64*, 73.

JENSEN, D. D., 1965, Anim. Behav. *13* (supplement), 9.

JOHN, E. R., 1964, Studies on learning and retention in planaria. *In:* M. A. Brazier, ed., Brain function, vol. II. Univ. Calif. Press, p. 161.

KIMMEL, H. D. and V. L. HARRELL, 1964, Psychon. Sci. *1*, 227.

KIMMEL, H. D. and R. M. YAREMKO, 1966, J. Comp. Physiol. Psychol. *61*, 299.

MCCONNELL, J. V., 1962, J. Neuropsychiat. *3* (supplement), 542.

MCCONNELL, J. V., 1966a, Ann. Rev. Physiol. *28*, 107.

MCCONNELL, J. V., 1966b, New evidence for the 'transfer of training' effects in planarians. Paper presented at XVIII international congress of psychology, symposium on the biological bases of memory traces, Moscow.

MCCONNELL, J. V., 1967, Specific factors influencing planarian behavior. *In:* W. C. Corning and S. Ratner, eds., Chemistry of learning; invertebrate research. Plenum Press, New York, p. 217.

MCCONNELL, J. V., A. L. JACOBSON and D. P. KIMBLE, 1959, J. Comp. Physiol. Psychol. *52*, 1.

THOMPSON, R. and J. V. MCCONNELL, 1955, J. Comp. Physiol. Psychol. *48*, 65.

VATTANO, F. J. and J. W. HULLETT, 1964, Psychon. Sci. *1*, 331.

WARREN, J. M. 1965, Ann. Rev. Psychol. *16*, 95.

WESTERMAN, R. A. , 1963, Science *140*, 676.

ZELMAN, A., L. KABAT, R. JACOBSON and J. V. MCCONNELL, 1963. Worm Runner's Digest *5*, 14.

CHAPTER 13

Transfer of behavioral bias: reality and specificity*

JAMES A. DYAL
Department of Psychology, University of Waterloo
Waterloo, Ontario

13.1. The context of the controversy

In 1965 a rather unprecedented event occurred. There were published almost simultaneously the results of four completely independent investigations which reported that 'learning' or 'memory' of 'behavioral bias' could be transferred from a trained rat to an untrained rat by injections of brain homogenates or extracts. These experiments had been stimulated by the controversial findings by McConnell and his associates that memory could be transferred from one planarian to another through the use of at least three techniques: regeneration (McConnell et al. 1959), cannibalism (McConnell 1962), and injection of RNA extracts (Jacobson et al. 1966b). Although the immediate stimulus for the experiments on interanimal transfer in mammals had been McConnell's work with planaria, this research, if viewed in a broader context, may be seen as one facet of the current effort in the 2.000-year-old search for the physical basis of the mind. You will recall that Aristotle had placed the mind in the heart; and a few hundred years later a big step was taken by Galen who took the mind out of the heart and put it in the head. The faculty psychologists and Gall and Spurzheim, the founders of phrenology, encouraged us to think that a whole host of psychological

* Early drafts of portions of the present paper were presented at the 1969 meetings of the Southwestern and Western Psychological Associations and at the symposium on 'Biology of Memory' sponsored by the Biological Sciences section of the Hungarian Academy of Science at Tihany, Hungary, September 1–4, 1969. The research and communication of the results have been supported by the Texas Christian University Research Foundation, University of Waterloo Research Grant No. 037-8737-02 and Grant No. APA 0305 from the National Research Council.

functions might be rather precisely localized in the brain. Florens, using the method of extirpation, and Fritsch and Hitzig, using direct electrical stimulation of the cortex, provided experimental evidence supporting the general notion of localization of the mind and specified some of the details. Franz and Lashley continued the use of extirpation as a technique and added more sophisticated behavioral measures, but after more than 30 years of intensive research Lashley was forced to conclude that his research program had 'yielded a good bit of information about what and where the memory trace is not. It has discovered nothing directly of the real nature of the engram. I sometimes feel in reviewing the evidence on localization of the memory trace, that the necessary conclusion is that learning is just not possible' (Lashley 1950).

Still the search continued. The last two decades have seen important and impressive developments along two relevant lines. First, the development of microelectrode techniques has permitted lesioning, stimulation and recording from smaller and smaller areas in the nervous system. Although a vast amount of new information has been generated by these techniques the engram has escaped the probe of the microelectrode. Second, the unprecedented breakthrough by molecular biologists in formulating and testing the powerful Watson–Crick model for the coding of genetic information encouraged neuroscientists to speculate that experiential memory might be coded in neuronal macromolecules.*

How molecules go about registering and storing information resulting from experience is still an intriguing mystery. Several different approaches have been taken toward the solution of this problem; and these approaches provided at least circumstantial evidence that indeed molecular registering and storage of experience does occur. We may distinguish at least three general approaches to unmasking the macromolecular identity of the engram. The first general approach has been to analyze changes in brain chemistry

* This proposition was first made by Katz and Halstead in 1950, thus amazingly antedating by three years the publication of the Watson–Crick model. However, McGaugh (1967) has pointed out that the idea was not a new one since about 50 years before Katz and Halstead, William James had written the following: 'Every smallest stroke of virtue or of vice leaves its never so little scar. The drunken Rip Van Winkle, in Jefferson's play, excuses himself for every fresh dereliction by saying, 'I won't count this time.' Well! He may not count it; but it is being counted nonetheless. Down among his nerve cells and fibres the molecules are counting it, registering and storing it up to be used against him when the next temptation comes'.

which are associated with differential experience. Several examples of this approach come quickly to mind. Krech et al. (1962) have shown in the laboratory rat that both brain chemistry and cortical structure are affected by experience. Experiments by Applewhite and Gardner (1968) and by Corning and Freed (1968) have shown that in protozoa and planaria, respectively, RNA synthesis reaches a peak and then drops off during the course of habituation and classical conditioning. Gaito et al. (1968) have shown differences in RNA and DNA ratios between trained and untrained animals in the ventrimedial cortex; Gaito has also used RNA hybridization techniques to provide suggestive evidence that unique species of RNA are manufactured in the brain during the learning of a shock avoidance task (Machlus and Gaito 1968).

More direct evidence supporting the involvement of RNA synthesis in the learning process have come from elaborate microdissection and microanalytic techniques of Hydén and his coworkers (see Hydén and Lange 1965). Furthermore, Zemp et al. (1966) have used double pulse labeling techniques to show that trained mouse brains incorporate significantly more radioactive uridine than appropriate untrained controls. One plausible interpretation of this effect is that dramatic increases in RNA synthesis occur during learning.

The second general approach is to attempt to facilitate or to inhibit synthesis of RNA or protein by injecting appropriate chemicals. For example, there are the attempts by Cook et al. (1963) to show that administration of yeast RNA results in the facilitation of learning. Results of experiments which have attempted to interfere with RNA synthesis through administration of 8-azaguanine (Dingman and Sporn 1961) or ribonuclease (Corning and John 1961) are consistent with the notion that decreased RNA synthesis interferes with learning and retention. Furthermore, Hendrickson and Kimble (1968) and Krylov et al. (1965) have shown that cortical application of RNAse results in poor retention. Interference with protein synthesis through injections of puromycin (Flexner et al. 1966), actinomycin D (Cohen and Barondes 1966) and cycloheximide (Barondes and Cohen 1967) have suggested that short term and long term memory are differentially affected by protein synthesis. The appropriate interpretation of the results from the application of all of these different techniques has generated conflict and controversy. However, it is quite clear that the third general approach – that of interanimal transfer of memory by injection – has generated by far the greatest controversy.

13.2. The original experiments

Reiniš undoubtedly made the first public statement of successful memory transfer between mammals in a paper presented at the 1964 conference for the experimental and clinical study of higher nervous functions (Reiniš 1965). Reiniš trained rats to approach a food cup with light and tone as the discriminative stimuli. He found that recipients injected with brain homogenates from trained donors found the food faster on the first test trial and had a shorter average response latency in the first test session than recipients injected with untrained brain homogenate or liver homogenate.

Ungar's first experiments demonstrated that both morphine tolerance and sound habituation could be transferred to naive recipients (Ungar 1965; Ungar and Oceguera-Navarro 1965). In the sound habituation experiments both rats and mice were habituated to the sound of a hammer hitting a steel plate. After the donors were thoroughly habituated their brains were removed and either a homogenate or a dialyzate was injected into the recipients, who were then habituated at 100 trials per day to the sound stimulus. A control group of recipients was injected with homogenate from non-habituated rat brains. These recipients were then given 14 days of habituation tests. The experimental recipients reached a 50% level of habituation on the very first day while the control group had not reached that level even after 10 days of testing. A highly significant and dramatic facilitation of habituation had resulted from the injection of '... peptide-type material extracted from the brain of habituated animals'.

The Copenhagen group trained rats in a light–dark discrimination apparatus to select the lighted side. The dependent variable was errors in percent of reinforcements. The differences between the group injected intracisternally with 'conditioned RNA' and the groups with 'unconditioned RNA' (or not injected), were significant when all test sessions were combined (Fjerdingstad et al. 1965).

The experiment by Jacobson's group was published in Science on August 6, 1965, and was the paper around which most of the early controversy centered (Babich et al. 1965b). They trained rats to approach a food cup to a discriminative click stimulus. Following training, a phenol extract rich in RNA was prepared from the brains of the donors and injected into naive recipients who were then tested by being placed in the Skinner box and presented with the click stimulus a total of 25 times. The number of approaches to the food cup within 5 sec was counted. The recipients injected with trained RNA made significantly more approach responses than control subjects injected with untrained RNA. The experiment was repeated using hamsters as donors and rats as recipients and yielded essentially the same result (Babich et al. 1965a).

These pioneering experiments confronted neuroscientists with three basic questions concerning the 'memory transfer' phenomenon. First, is it real? Second, if it is real (i.e. experimentally reproducible), how specific is it to the original training conditions? Third, what are the chemical characteristics of the active transfer agents? As the title of this paper implies, the remainder of the discussion will deal with the first two questions.

13.3. The question of reality

Immediately after the publication of initial positive reports in 1965, there ensued a period of intense research aimed at determining the reliability of the phenomenon. The first negative attempt to replicate the Jacobson results was reported by Gross and Carey (1965); similarly, Gordon et al. (1966) failed to replicate Jacobson. Luttges et al. (1966) reported a series of eight experiments using a variety of tests (none of them like the Jacobson test); they obtained uniformly negative results. In one of the experiments they labeled brain RNA in vivo with ^{32}P. They then extracted this labeled RNA, injected it i.p. into recipient rats, and measured the amount of radioactivity in the brain; they found no evidence that the RNA passed the blood–brain barrier. Not unreasonably this experiment convinced many people that the memory transfer effect could not be relevant to learning even if it were reproducible. Reproducibility appeared to be much in doubt when there was published in Science a letter signed by some highly distinguished scientists who reported that they had conducted a total of 18 experiments on memory transfer and had obtained consistently negative results (Byrne et al. 1966). This report seemed to signal the coup de grace for the phenomenon of memory transfer for the general scientific community, since few readers gave sufficient weight to the last paragraph of the letter which reads as follows:

'Our consistently negative findings do not of course bear directly on the possibility that RNA may be involved in the mechanism of memory. They indicate only that results obtained with one method of evaluating this possibility are not uniformly positive. Furthermore, we feel that it would be unfortunate if these negative findings were to be taken as a signal for abandoning the pursuit of a result of enormous potential significance. This is especially so in the light of several other related but not identical experiments that support the possibility of transfer of learning by injection of brain extract from trained donors. Failure to reproduce results is not, after all, unusual in the early phase of research when all relevant variables are as yet unspecified.'

Indeed, a great deal of other related work was being conducted in other laboratories which seemed to '...support the possibility of transfer of learning by injection of brain extract from trained donors'.

13.3.1. Rosenblatt's research

In 1966, Rosenblatt reported two positive replications of Jacobson's original experiment (Rosenblatt et al. 1966a). The Cornell laboratory then reported a series of experiments which represent a most massive attack on the problem. The first 10 experiments involved some 242 donors and 487 recipients using a variety of learning tasks and a variety of extraction procedures (Rosenblatt et al. 1966b, c). Statistically significant effects were not obtained with every extract or training procedure. However, when one considers all the data reported, significant positive memory transfer was demonstrated for an operant response in a Skinner box, for active avoidance in a two-way shuttle box, for passive avoidance in a shuttle box, and for a compound shuttle-bar press response when 'light-on' was the positive stimulus.

The next series of 10 experiments involved over 600 donors and 300 recipients conducted on a variety of left–right discrimination tasks, e.g. T-maze, Y-maze, two-bar boxes and double panel-pushing boxes (Rosenblatt and Miller 1966a, b). The data analyses were done on two types of dependent variables: (1) activity measures (e.g. in the bar boxes total number of responses whether correct or not was taken as an index of activity, and in the T-maze response latencies were used as an activity measure); and (2) a differential responding score which was the number of responses to the left made by the left-injected recipients minus the number of left responses made by the right-injected recipients.

The results using the activity measures as the dependent variable were uniformly negative; there was no transfer based on activity indices. Analysis of the differential response score failed to demonstrate significant transfer in the T-maze situation. When all of the data for the various extraction procedures were combined a significant effect was obtained for the two-lever box situation. Furthermore, a highly significant transfer effect was obtained with the two-panel box when the injection was an acetone precipitate of a homogenate supernatant injected i.v. However, there were no significant effects for any other extracts or for i.p. injections. Pooling all of the data from the two-bar and the two-panel experiments for all extraction procedures and all injection routes overall significant positive transfer effects were obtained. It should be kept in mind that all of the work reported by Rosenblatt in 1966 was exploratory in the sense that he was manipulating a large number of variables hoping to come upon some procedures which would give positive results; thus in the present instance the combined probability results from the combining many null result experiments with a few procedures that yielded positive results and is thus a very conservative test.

Rosenblatt and Miller's results in the Y-maze are complicated by the fact that both significant positive effects and significant inversion effects were obtained, and the direction of the effect appeared to be dependent on the dosage level. Thus, from Rosenblatt's early work it seemed reasonable to conclude that the effect depended upon complex interactions of behavioral variables (e.g. type of task: T-maze versus Y-maze), physiological variables (e.g. injection route: i.p. versus i.v.), and biochemical variables (e.g. dosage and type of extract). Perhaps it is not too surprising that a large number of negative results was obtained in the early experiments (see Byrne et al. 1966).

13.3.2. McConnell's research

McConnell began his studies of interanimal transfer in mammals in 1965 with a successful replication of Jacobson's first experiment. He found that subjects injected with trained RNA made more cup responses than those injected with untrained RNA or trained RNA which had been treated with ribonuclease. The second procedure used by McConnell and his associates (McConnell et al. 1969) involved the transfer of the facilitating effects of dipper training in a Skinner box. The experimental donors were trained to approach and drink following the distinctive click made by the dipper when it was elevated into the chamber. The control donors were not trained. Following injection all recipients were trained to bar press for milk reward. The experimental question was whether or not the previous dipper training of the experimental donors would facilitate learning of the bar-press response. McConnell conducted eight of these experiments, seven of which gave significant positive transfer effects. In addition, he demonstrated again in two of the experiments that the effect was eliminated when the trained RNA was incubated with ribonuclease. Furthermore, he showed that the size of the transfer effect depended upon the amount of training of the donors. Injections from experimental donors which were trained for 10 days had a considerably greater facilitating effect than those from animals that had been trained only five days.

13.3.3. Ungar's research

After demonstrating transfer of habituation and of morphine tolerance (Ungar and Oceguera-Navarro 1965; Ungar and Cohen 1966) Ungar's next series of experiments utilized escape in a Y-maze. In the first three experiments the donors were trained to escape into the lighted arm. The recipients were tested every day for five days after injection. They were given 20 test trials per day with no shock in the maze arms to see whether they turned toward the lighted arm or the unlighted arm. In each of the experiments the recipients of trained brain showed a significant increase over their preinjection level in number of turns to the lighted arm. The control animals which received naive brain injections showed no such increase (Ungar 1967a).

These experiments were followed by three experiments in which the donor rats were trained to go either left or right to escape shock (Ungar and Irwin 1967). The recipients were tested for five days following injection. On days 1 and 2, the number of left turns from left-injected recipients was significantly higher than their preinjection level. The right-injected subjects were significantly different from their preinjection level on the third day following injection. The untrained-injected subjects were not significantly different from preinjection level on any of the days. If one combines the data for all three of these left–right experiments and simply counts the changes in direction of turn from pre- to postinjection, it is found that 57 of the animals injected from donors trained to go left, made more turns to the left following injection than they did prior to injection; whereas, 32 of them made more turns to the right. On the other hand, of those that were injected with right turn homogenate, 53 made more turns to the right and 35 more turns to the left. The χ^2 for the resulting four-fold table has a probability of less than 5 in 1000.

Ungar's next series of experiments involved passive avoidance learning in a modified version of a task first used by Gay and Raphelson (1967). The three-compartment apparatus consisted of a start box with two alleys on either side leading to either white or black boxes. Capitalizing on the fact that rats normally prefer the black box, Gay and Raphelson proceeded to shock the donors in the black box, inject the extracts into recipients and then test the recipients to see if this changed the relative amount of time they spent in the two boxes. They found that the mean amount of time spent in the black box by the experimental group out of 3 min was 15 sec, whereas the control group spent 120 sec.

Ungar felt that this was a particularly appropriate behavioral assay with which to demonstrate the memory transfer effect and in 1967 and 1968 he tested over 800 recipients in this passive avoidance task. Over all of these experiments the average time spent in the dark box by the recipients of the trained brain was 63 sec and by the controls 118 sec. The probability associated with this difference was less than 1 in 5000 (Ungar 1968).

Ungar has also obtained positive transfer effects using a step-down box. Prior to injection the step-down latency of the experimental and control subjects were equivalent. Following injection, the experimental animals took significantly longer to step down from the platform than do control animals (Ungar 1969).

13.3.4. Dyal's research

In January of 1966, I became interested in the memory transfer field, and together with Arnold Golub and Robert Marrone began a series of experiments designed to provide further evidence on the replicability of the phenomenon first reported by Jacobson's group. The subjects in the first experiment were 36 Sprague–Dawley rats between 100 and 150 days old. Twenty-four of them were assigned at random to three donor groups and the remaining 12 subjects were assigned to three recipient groups. Group A was given 10 days of continuously reinforced bar press training (CRF) in a Grayson–Staedler Model E 3125 Skinner box slightly modified by moving the bar to the side opposite the food magazine and by passing a photoelectric beam across the entrance to the food cup. When a rat stuck his nose into the food cup, the circuit was broken and the event was automatically counted thus eliminating the possibility of experimenter bias in counting the critical response. Depression of the bar resulted in the occurrence of a compound event consisting of a click of the food delivery mechanism, the delivery of the food pellet (Noyes 45 mg), and the diminution of the light in the photoelectric circuit. The second group (group A-Y) was yoked to the group A subjects so that whenever a group A rat pressed the bar the 'light–click' occurred in the yoked control box but the food pellet was not delivered.

When a control subject pressed the bar, the event was recorded on a counter but the bar did not activate the food delivery mechanism. The third group of donors received 10 days of CRF training followed by three days of extinction, followed in turn by three days of reacquisition on a CRF schedule; this group will be designated as group AER (acquisition-extinction-reacquisition).

Within 1 hr following completion of the last training session the donors were killed with ether. The brain, excluding the olfactory bulbs and the cerebellum, was removed within 2 min of expiration, placed into a 25 ml tissue grinder and stored in a refrigerator at 0 °F. The brains were homogenized with 1 cc per brain of physiological saline, and 3.2 cc of the resulting homogenate was injected i.p. into the recipients who were tested 24 hrs later. They were 48–56 hrs food-deprived at the time of test. The testing was conducted in two sessions each of 30 min duration. During the first test session the bar was removed from the box and the 'light-click' stimulus was presented automatically every 60 sec but a food pellet was not delivered. The number of times that each rat stuck his nose into the food magazine (magazine entry, ME) was automatically recorded. During the second test session the bar was activated on a CRF schedule and both bar presses and magazine entries were recorded. Only the first test session proved to be consistently sensitive to transfer effects thus the data are described in terms of this dependent variable. The results of the first experiment may be seen in fig. 13.1. Group AER exhibited a strong transfer effect making roughly twice as many magazine entries as the yoked control group. Statistical comparison of the two groups using a Mann–Whitney U test resulted in a U value of zero, that is, there was no overlap between the two distributions. The associated p value was 0.01 ($n_{1,2}=4$). No evidence for transfer was obtained in group A. We did not understand why a transfer effect should have been obtained with group AER but not with group A and proceeded to determine if the effect obtained with group AER was reliable.

The procedures in experiment 2 were the same as in experiment 1 with the exception that only group AER and its yoked control (AER-Y) were run; there were 12 donors and six recipients per group. As may be seen in fig. 13.1, the results were essentially the same as experiment 1. The experimental group again made approximately twice as many magazine entries as did the control group ($U=5$; $p=0.04$). Further details on these two experiments may be found in Dyal et al. (1967).

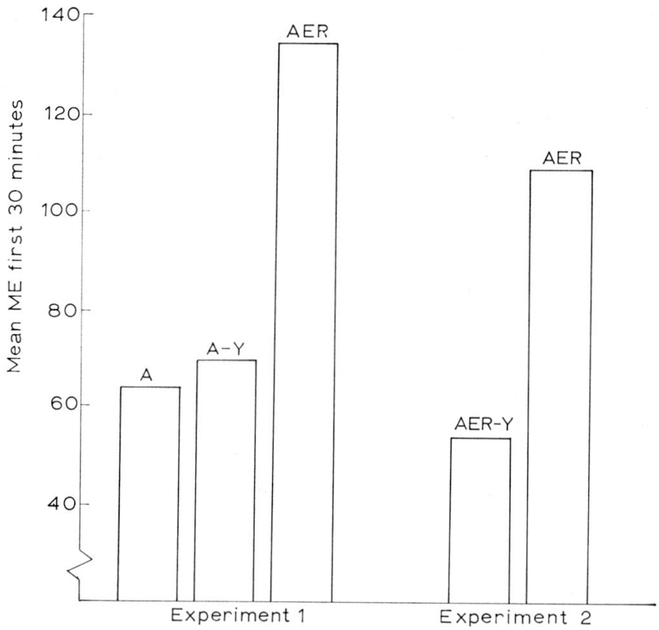

Fig. 13.1. Mean number of ME's made by experimental and control recipients during the first 30 min test in experiments 1 and 2.

Recognizing that statistical reliability does not guarantee experimental reproducibility, Golub and I proceeded with two further replications of this paradigm (Dyal and Golub 1968). In our third replication the procedures were essentially the same as in experiments 1 and 2 except that preinjection operant levels of magazine entry were measured under the same conditions which were present during the postinjection tests. Each recipient group contained six subjects. As may be inferred from inspection of fig. 13.2, there were no reliable increases in response rate from pre- to postinjection scores for the experimental group (AER) or the yoked control group. The saline group on the other hand made significantly more magazine entries on their postinjection test. Experiment 3 had been conducted with relatively untrained undergraduate research assistants, and an excessive number of procedural errors and equipment failures occurred in the training of the donors. We repeated the experiment with the same experimenters who were now somewhat better trained. As may be seen in fig. 13.2 the results of the fourth replication are quite in line with experiments 1 and 2. Both groups increased their magazine entries on the postinjection test but the experimental recipients increased reliably more than the control recipients (U=8; p=0.037). The procedure recommended by Winer (1962) for determining the combined probability of a series of experiments conducted to test the same hypothesis was applied to the results of these four experiments. The resulting χ^2 value of 25.43 was significant beyond the 0.001 level.

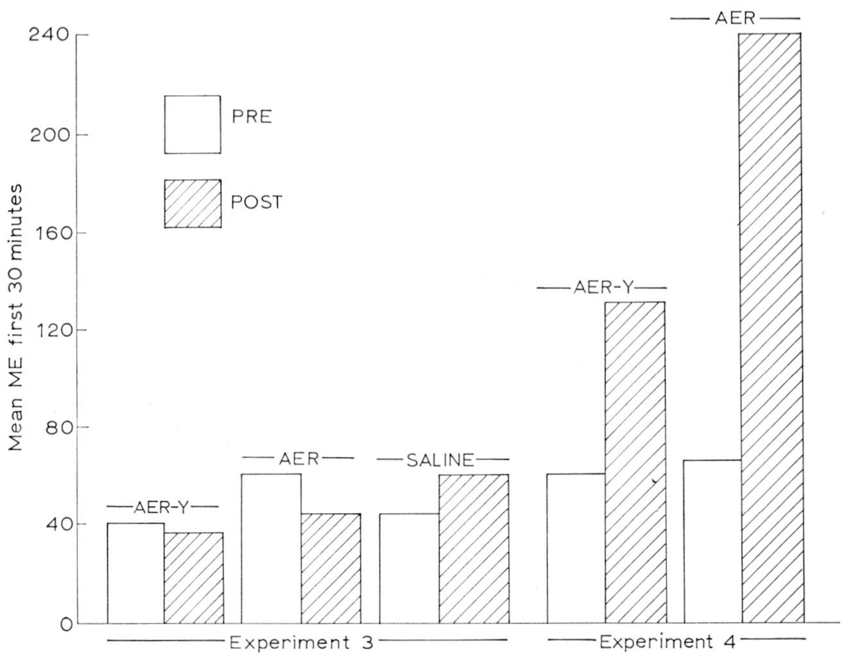

Fig. 13.2. Mean number of ME's made by experimental and control groups prior to injection and during the first 30 min test in experiments 3 and 4.

At this point in the research we felt reasonably confident that this procedure could be used, by us at least, to obtain positive transfer effects, so we proceeded to work on the problem of response specificity. This experiment will be discussed later; suffice to say that the apparatus was modified to accommodate the special needs of this experiment. Shortly thereafter both Golub and I left our TCU laboratory for postdoctoral work in other laboratories. The research program at TCU was assumed by Dr. Wayne Ludvigson who attempted two replications of our transfer paradigm. In the first of the Ludvigson replications 148 male rats of the Simonson strain were assigned at random to four donor groups of 20 sessions each and four recipient groups of 12 sessions each. The four donor groups consisted of the three treatment groups of the previous experiments (A, AER and AER-Y) plus a fourth yoked control group (AER-Y^{+}) which differed from AER-Y in that it received food pellets plus 'light–click' when the AER group pressed the bar. However, the bar in the AER-Y^{+} box was not activated. The AER-Y^{+} group thus served to determine whether the operant response of bar pressing was required for transfer or whether the critical event was merely the contingency between 'light–click' and the receipt of food. The results of the experiment indicated no significant differences between any of the groups on either of the dependent variables (bar press or magazine entry) on either the first 30 min test period, the second 30 min test period, or for five successive days in which learning enhancement in barpress acquisition was tested. Again, the difficult questions associated with null results arise. In the present instance the experiment was conducted by two new untrained undergraduates; 21 procedural errors or apparatus failures occurred during the training of the donors. Furthermore, the animals were of a different strain and supplier. Although these results must be reported as a failure to replicate the previous positive findings, the reason for the failure is unclear. It should be noted that in a research area less controversial than this one the failure to reproduce an effect when there were serious reasons to question the adequacy of the procedures, would typically not even be reported!

The second attempted replication by Ludvigson (Dyal et al. 1969) used the same treatment groups as in the first replication with an additional non-injected group. There were 12 recipients in each group. The homogenization and injection procedures were the same as in the first four experiments. The results are presented in table 13.1 for the magazine entry measure. Analysis of the pre- versus postinjection difference scores revealed that all of the groups in which the donors had the 'light-click' stimulus paired with food (A, AER, Y^{+}) made significantly more magazine entries than the yoked control group (Y^{-}) in which the donors received no food with presentation of the 'light–click' (p at least 0.05 for a two-tailed test).

Although this difference score analysis would suggest that a positive transfer effect was obtained, closer scrutiny reveals that this is not the case. First, the difference score performance for the Y^{-} control group is also significantly below that of the non-injected (NI) control group. Thus the difference between Y^- and all groups could be due to a significant suppression of performance of Y^- rather than an elevation of performance in the other groups. However, as may be seen in table 13.1, all the above group differences based on difference score data are due to the fact that the performance of Y^- was elevated over that of the other groups on the preinjection baseline. Thus, meaningful group comparisons cannot be based on the pre versus post difference scores but could be based on postinjection scores per se*. Examination of table 13.1 reveals that those groups of recipients whose donors had the 'light–click' stimulus paired with food made more magazine entries than either of the control groups (Y^- and NI); however, analysis of these postinjection scores by means of the Mann–Whitney U test resulted in no statistically reliable differences between any of the groups.

TABLE 13.1

Mean number of magazine entries made during preinjection baseline tests and postinjection transfer test.

Treatment	Preinjection	Postinjection	Mean difference score
A	64.00	94.08	30.08
AER	74.08	82.00	7.92
Y^+	49.75	81.75	32.00
Y^-	106.58	64.83	−41.75
NI	59.33	62.75	3.42

The mean number of bar presses made in the first 30 min test by each group is presented in table 13.2. On this dependent measure group AER made reliably more postinjection bar presses than the Y^- control group; however, group AER was not significantly different from the other two control groups (Y^+ and NI). Similarly group A was not reliably different from the three control groups.

Another analysis applied to these data rests on an argument made by Rosenblatt: '...we have noted that the transfer effect seems to 'take' on some rats while leaving others unaffected... In cases where a successful transfer

* In contrast to expectation the correlation between test 1 and test 2 for the non-injected group was +0.13; thus the postinjection performance of the experimental animals may be relatively un-influenced by preinjection performance level.

TABLE 13.2

Mean number of bar presses made during the preinjection baseline tests and the postinjection transfer test.

Treatment	Preinjection	Postinjection[a]	Mean difference score
A	4.50	9.38 (11.17)	+4.88
AER	3.50	8.37 (8.17)	+4.87
Y^+	3.38	2.50 (3.83)	−0.87
Y^-	6.38	3.75 (3.82)	−2.63
NI	6.00	4.37 (6.17)	−1.63

[a] Preinjection baselines were not obtained on four of the recipients in each group, therefore the scores entered under Preinjection and Postinjection are for $n = 8$. The postinjection scores in parenthesis are the mean scores for all 12 recipients in each group.

seems to have occurred, the experimental group generally has a number of extremely high scores, below which is a middle region consisting predominantly of controls and at the bottom an indistinguishable mixture of control and experimental scores' (Rosenblatt et al. 1966c, page 553).

We also have found that the data from our successful experiments are distributed in this manner, and thus have proceeded to analyze only the top 50% of the scores in each distribution as a potentially more sensitive test of marginal effects. The results of this analysis support the significance of the pattern of results which is present in table 13.2, namely the two experimental groups receiving bar-press trained homogenate (A and AER combined) made significantly more bar presses than comparable control groups receiving brain homogenates from animals which had not been trained to bar press (Y^+ and Y^- combined; $U=29$, $n_1=n_2=12$, $p<0.02$).

The overall characterization of the results of the second Ludvigson experiment would emphasize that the pattern of the group differences was as would be expected if positive memory transfer had occurred. Furthermore, the data suggest that when magazine entry is the dependent variable the critical event is simply that the 'light–click' stimulus be reinforced by pairing with food delivery (group Y^+ tended to perform more like groups A and AER than group Y^-). On the other hand when bar pressing is the dependent variable, bar press training is necessary for transfer (groups A and AER perform similar to each other and make more bar presses than Y^+ and Y^- which tend to be similar to each other). However, the fact that these generalizations rest on quite marginal effects statistically and may tend to capitalize on chance differences in multiple group comparisons requires that they be held most tentatively. When both experiments are considered, it is clear that the Ludvigson attempts at replication of the Dyal and Golub results were not strongly supportive; however, it is unclear how much changes in the procedure and apparatus may have effected these results. It should be noted that Golub and McConnell have replicated the basic Dyal and Golub procedures in several experiments at the University of Michigan (Golub and McConnell 1968; Golub and McConnell 1969). They have repeatedly obtained learning enhancement of bar press acquisition as a result of injection of homogenate or RNA extracts from AER trained donors.

Considering the whole set of experiments which have utilized the AER training procedure in a single bar box it seems reasonable to conclude that the positive transfer effect is not only statistically reliable but also has some meaningful degree of translaboratory replicability.

Following the basic demonstration of the positive transfer effect using an AER procedure, our interests turned to the question of specificity. However, even in the process of devising transfer procedures which might be especially relevant to problems of specificity we were of necessity also concerned with being able to obtain the basic transfer effect. Thus, in collaboration with Dr. James Cornell an experiment was conducted to determine if a conditioned fear procedure previously utilized by Palfai and Cornell (1969) could provide a sensitive behavioral assay for transfer of fear conditioning. Palfai and Cornell had shown that if rats were given several pairings of a tone and shock in one enclosure and were then tested in an open field with tone alone they tended to 'freeze' to the tone and that this 'freezing' measure seemed to be sensitive to the amount of fear which had been conditioned. The Palfai and Cornell procedure was applied to the transfer problem in the following manner:

Forty-seven male Sprague–Dawley rats were trained in a Grayson–Stadler shock box under one of three treatments: classical fear conditioning (CC); foot-shock only (FS), and untrained (U). For the classical conditioning group the conditioned stimulus was a 1000 Hz, 80 db tone of 60 sec duration; its offset coincided with the onset of a 0.1 sec duration, 2 mA foot shock. The FS group received an equal amount of shock but no conditioned stimulus. The untrained group was put into the box an equal amount of time. All sessions were trained 15 trials per day for five days; the intertrial interval ranged from 30–120 sec ($\bar{X}=60$ sec). The brains were removed 48 hrs after the last training session (a two-day 'incubation' period). Brain homogenates were injected i.p. in the amount of 2.5 cc (1.5 brain equivalents).

The recipients were tested for four days in a 12 × 12 × 14 in plywood box with a wire mesh floor. A single test session was divided into four 3-min periods: adaptation, preconditioned stimulus, conditioned stimulus and post-conditioned stimulus. The duration of freezing behavior was measured during each of the latter three periods. In addition to the three homogenate recipient groups a saline group was run.

When all test sessions were combined it was found that the fear conditioned group (CC) froze significantly longer than the untrained group ($z=1.69$; $p=0.045$); however it did not differ from foot-shock-only group. On the third day following injection the difference between the experimental group and the combined control groups was highly significant ($z=2.20$; $p=0.014$).

The top 50% of the scores from each group were also compared. This analysis revealed a similar picture for all test days combined; while on day 3 the experimental group froze reliable more than either of the control groups individually or combined.

In summary, when all four test days are combined, the conditioned group (CC) was not reliably discriminable from the foot-shock (FS) control group. However, the CC group tended to freeze longer than the untrained group whereas the FS group did not freeze reliably more than the untrained group. Furthermore, on the third day of testing transfer of a *conditioned* fear was demonstrated in that the CC group froze reliably more than the two control groups (FS and U) combined. While it may be concluded that a conditioned fear was transferred via injection of brain homogenates, the experiment also emphasizes the importance of controls for sensitization effects. Experiments purporting to demonstrate transfer of conditioned fear by comparing conditioned groups with untrained controls should not be taken as valid. Nonetheless, it should be noted that the transfer of sensitization processes involved in the foot-shock-only treatment is interesting in its own right and would merit research aimed toward determining the active biochemical agent.

Golub et al. (1970) have suggested that the introduction of rest periods of two to five days during the training of donors seems to consolidate or incubate the learning which took place prior to the rest and results in a facilitation of the transfer effect. Also, in an unpublished paper, Golub and McConnell (1969) reported that introduction of a five-day rest period in the middle of a six-day training period results in a significant increase in resistance to extinction of the donors compared with control animals not receiving the rest period. Both of these findings are quite interesting and deserve further research; it would be especially important to show in a single experiment the occurrence of both 'incubation of learning' (as evidenced by increased resistance to extinction in the donors) and 'learning enhancement' (as evidenced by more rapid learning of recipients which were injected with the brains of 'incubated' donors). In the summer of 1969, I conducted an experiment to test the replicability of both effects. Twenty-four Sprague–Dawley rats approximately 80 days old were assigned at random to two treatment conditions. Group $A_4R_5A_2E_5$ was given four days of bar press training on a CRF schedule, then five days of rest followed by two more days of CRF training followed in turn by five days of extinction. Group $A_6E_3R_5E_2$ was given six days of CRF training followed by three days of extinction, then five days of rest and two more days of extinction. Considering the procedure only through the first three days of extinction, these two treatments are identical to the groups which showed the greatest differences in resistance to extinction in the Golub and McConnell (1969) incubation-of-learning study.

Analysis of the number of bar presses on the first day of extinction revealed that when acquisition level was taken into account by the use of difference scores, group $A_4R_5A_2E_5$ extinguished significantly more rapidly than group $A_6E_3R_5E_2$. These results are the opposite of those reported by Golub and McConnell.

After the last extinction session brain homogenates were prepared and injected i.p. in the amount of 1.8 brain equivalents into naive recipient rats. The recipient groups received injections from either the $A_4R_5A_2E_5$ group, the $A_6E_3R_5E_2$ group, or they were injected with an equal volume (4.0 cc) of saline. If incubation effects were critical determiners of the transfer phenomenon, recipients injected with $A_4R_5A_2E_5$ brains would show learning enhancement when trained under a CRF schedule, in comparison with saline controls. The $A_6E_3R_5E_2$ group would be expected to be less facilitated than $A_4R_5A_2E_5$, or if transfer of extinction occurs, to be even inhibited relative to saline controls. The recipients were 72 hrs food-deprived during the first training session, which occurred 24 hrs after injection. No differences between the recipient groups appeared on any of the five days of training following injection (Dyal 1969).

13.3.5. Positive demonstrations from other laboratories

Further research from Reiniš' laboratory supports the reliability of his conditioned approach response procedure. In all the experiments the subjects were Swiss mice trained to open a plexiglass door when a light came on in order to obtain food. Reiniš and Kolousek (1968) reported that recipients injected with trained extracts made reliably more approach responses in the first test session than control groups which were either not injected or injected with brain extracts from untrained donors. Similar positive transfer effects have been reported in other experiments from this laboratory (Reiniš 1968, 1969).

Following his initial series of unsuccessful experiments (Byrne et al. 1966) Byrne has reported several experiments in which positive transfer effects were obtained. Byrne and Samuel (1966) reported an experiment utilizing a two-bar operant chamber with procedures patterned after those of Rosenblatt et al. (1966b, c). They obtained an asymmetrical transfer effect in that the recipients of extracts from donors trained on the right bar learned significantly faster than recipients of untrained brain, however, the recipients of left-trained brain extract showed no such transfer. Further experiments with a mirror-image Skinner box suggested that the critical variables might be the relative discriminability of the two bars in relation to the chamber door (Byrne and Hughes 1967).

The Copenhagen team has collaborated with Byrne to compare the results obtained with different extraction and injection procedures in the same behavioral situation. They obtained significant positive transfer of a left–right discrimination in a two-bar Skinner box with both a tris-NaCl extract (two brain equivalents injected i.p.) and with an 'RNA' phenol extraction and dialysis ($\frac{2}{3}$ brain equivalents injected intracisternally; Fjerdingstad et al. 1970). Fjerdingstad has continued his own research program and has demonstrated positive transfer of a learned preference; donors were reinforced with water for entering the white compartment in an Ungar-type (Ungar et al. 1968) black–white apparatus. The experimental recipients did not differ from the controls in time spent in the white box, but they did spend significantly less time in the black box. Fjerdingstad has also presented a series of five experiments which he interprets as indicating '...that when brain extracts from animals trained to alternate are injected into recipients, an increased tendency to alternate is caused' (Fjerdingstad 1969). This conclusion is based on pooling the data for five experiments over six test days. In the case of the individual experiments three yielded a statistically significant effect either on a given test day or when all test days were combined. Fjerdingstad has obtained positive transfer effects using goldfish as subjects in a shuttle-box active avoidance task. The details of these experiments are available in chapters 4 and 11 in this volume.

Faiszt and Ádám (1968) conducted a series of 12 experiments investigating the transfer of an active avoidance response. Donors were trained to jump up on a shelf in order to avoid foot shock. Recipients were typically trained to jump to the unconditioned stimulus only and then following injection of either trained or untrained RNA extracts they were trained to make an avoidance response to the conditioned stimulus. In seven of the 12 experiments the learning of the experimental recipients was significantly facilitated compared to the controls. Transfer did not occur when (1) the recipients were habituated to the conditioned stimulus prior to conditioning, (2) when they were 'overtrained', or (3) when they were young animals. Optimal transfer effects occurred when the donors were trained for 10 days (twelve 4-trial sessions) and when adult recipients were not habituated to the conditioned stimulus. Faiszt and Ádám (1968) have reported four additional positive replications of this procedure and have presented data to support the contention that the active factor in the transfer is in the ribosomal RNA.

Weiss (1970) trained rats to avoid drinking from the drinking tube mounted on their home cage when a small light mounted on the front of the cage was turned on. Following 48 hrs of continuous training all animals were discriminating perfectly, i.e. they drank only when the light was off. Training was continued for another 50 days. The donors were then sacrificed, and an RNA extract was prepared and injected i.p. into 10 recipients. The recipients were then tested with alternating 'light-on' and 'light-off' periods of 3 sec duration for a period of 48 hrs. The recipients of trained brain extracts made significantly fewer approaches to the drinking tube during the 'light-on' period than during the 'light-off' period, whereas a saline-injected control group showed no such differentiation. Wolthuis (1969) has used a procedure similar to that of Weiss except that rats were trained to drink during the 'light-on' period. He obtained highly significant transfer effects over three experiments. Furthermore, Wolthuis demonstrated significant positive transfer of a learned alternation pattern in a T-maze.

Several other laboratories have conducted successful transfer experiments. Daliers and coworkers have reported that habituation to audiogenic seizures may be facilitated by injections of brain extracts from animals which have previously had seizures (Daliers and Rigaux-Motquin 1968). Furthermore, Daliers and Giurgea (1969) have reported that spinal fixation time (i.e. minimum time needed for the spinal cord to maintain the postural asymmetry resulting from cerebellar lesions) is reduced by injections of brain extracts from animals which have been maintained in postural asymmetry. In addition to their previously noted research on incubation effects Golub and McConnell have utilized a single 'reminder' shock trial to demonstrate transfer of a passive avoidance response (Golub et al. 1969). They have also obtained positive transfer effects in a one-way active avoidance task (Malin et al. 1971). Kleban et al. (1968) have also reported what appears to be a positive transfer of an active avoidance task, although the possibility that the effect is due to transfer of differential activity cannot be excluded. Of perhaps more interest, their results suggest that extinction of an avoidance response in recipients parallels the resistance to extinction of the donors. Chapouthier and his colleagues have also obtained positive transfer of a simple avoidance under conditions which seem to preclude differential activity as the explanation (Chapouthier et al. 1969; Chapouthier and Ungerer 1969).

Garcia et al. (1955) showed that when rats are given whole body gamma irradiation following 30 min access to a saccharin solution they develop a strong aversion to saccharin. Previous attempts to transfer irradiation-induced conditioned saccharin avoidance (Revusky and DeVenuto 1967; McConnell et al. 1969) were promising but equivocal. However, very strong transfer effects have been reported recently by Moos et al. (1969). They subjected two groups of donor mice to whole-body gamma irradiation. A saccharin preference was established in the experimental donors while the control donors had access to tap water only. Injection intracerebrally of very small dosages (0.03 ml) of brain homogenate in Ringer's solution into naive recipients produced a very dramatic effect; whereas the control recipients increased their preference for saccharin over a 22-day test period, the experimental recipients exhibited a strong aversion to saccharin which lasted 22 days. The differences between the two groups in saccharin intake was highly significant ($p<0.001$).

13.3.6. Experiments reporting null effects

As previously indicated, early reports of memory transfer were followed by attempted replications in several other laboratories. Negative attempts to replicate the Jacobson procedure (Babich et al. 1965) were reported by Gross and Carey (1966), and Gordon et al. (1966). Albert (1966), in an elegant series of studies, had demonstrated intra-animal (interhemispheric) transfer effects via injections of brain RNA from a trained hemisphere into an untrained hemisphere. He was, however, unable to obtain interanimal transfer.

Some aspects of the experiments which served as a basis for the Byrne et al. (1966) letter in Science should be specified briefly. Leaf, Dutcher, Horovitz and Carlton attempted to transfer a CER as indexed by suppression of the rate of a drinking response during a stimulus which had been previously paired with shock (Leaf and Muller 1965). Their two experimental recipient groups received one or four brain equivalents of RNA extract. They obtained no differences in suppression between experimental and control recipients. Corson and Enesco attempted to replicate the original experiment by Babich et al. (1965b) in which experimental recipients were trained to approach a food cup to the sound of a click. They were unsuccessful in replicating the results or the experimental procedures. As they point out their procedures differed from those of Babich et al. in at least 13 ways; however, it would appear that most of these differences are 'trivial' and the reasons for the lack of confirmation are obscure. Corson and Enesco (1968) reported a series of eight experiments which attempted to transfer avoidance conditioning. Four of these experiments were patterned after Ungar's dark avoidance experiments in a Y-maze (Ungar 1967) and one was an attempt at an exact replication. None of the experiments resulted in positive transfer effects. Wagner et al. (1966) were unable to replicate transfer of a brightness discrimination reported by Jacobson et al. (1966a).

Miller et al. (1969) have been cognizant of the importance of *intra*-laboratory replicability of transfer effects and have conducted a series of five experiments using exactly the same behavioral procedures, which involved training rats to make a hurdle jumping avoidance response in a two compartment box previously described by Murphy and Miller (1955). The first three experiments were conducted with injection of RNA extracts whereas the last two were performed with brain homogenates. In the first experiment an inverse effect was obtained in which recipients of trained RNA learned significantly more slowly than those receiving material from handled controls. Experiments 2–5 obtained no reliable differences between experimental and control recipients.

Krech et al. (1967, see also chapter 8) reported several attempts to reactivate a previous experience through injections of 'reminder' extracts. They followed the procedure of Campbell and Jaynes (1966) in which weanling rats (21 days old) were given shock in a black box and permitted to rest in a white box. When these animals were tested four weeks later they showed complete forgetting of the original experience in that there was no avoidance of the black box. A second group was given the early traumatic experience plus one shock trial per week in the black box. These animals showed a strong preference for the white box whereas control animals given only the weekly 'reminders' without the original training showed no such preference. In the Krech et al. experiments weekly injections of brain homogenates from trained animals were used as the 'reminders'. Positive results were obtained in some of the experiments with brain homogenates and in one experiment with liver homogenate. Nonetheless, because of later difficulties in replicating these effects the results of the whole series must be regarded as equivocal. Bennett (personal communication) has independently conducted a large number of experiments similar to those reported by Byrne and Samuel (1966) but has generally been unable to obtain positive effects.

Attempts to transfer brightness and spatial discriminations in a variety of T-mazes or Y-maze choice situations have not generally been successful (Greene and Kimble 1966; Kimble and Kimble 1966; Rosenblatt and Miller 1966a; Wagner et al. 1966; Dyal and Golub 1967; Hoffman et al. 1967; Allen et al. 1969).

Several attempts have been made to replicate the transfer of a learned fear of a black box reported by Gay and Raphelson (1967) and by Ungar et al. (1968). Lagerspetz (1969) reported the results for three experiments involving a total of five recipient groups which were injected with homogenates from donors shocked in the dark box. Only one of these five groups showed a significant reduction in dark/light preference following injection. Golub and McConnell (1969) also failed in five of six experiments which attempted to replicate Ungar et al. (1968). Similarly Hutt and Elliott (1970) failed to replicate the positive transfer reported by Ungar et al. Although this experiment appears to have been a generally successful effort to repeat Ungar's procedures, it did deviate in that the donors were placed directly into the black box rather than being forced into it through the passageway. This deviation changes the task from a combination avoidance/fear conditioning task to a straight fear conditioning task. Only future research will tell if this difference is critical. Other researchers have failed to replicate Ungar's reported transfer of morphine tolerance (Tirri 1967; Smits and Takemori 1968); however, the replication procedures were not precise.

13.3.7. Conclusions regarding reality of the phenomenon

It is apparent that the pioneering experiments on memory transfer in vertebrates have generated a substantial amount of research since their publication in 1965. Considering that the area is still befuddled and beclouded by claims and counterclaims, by replications and failures to replicate, what, if anything, can be concluded regarding the reality of the phenomenon? In attempting to draw the most reasonable inference which can be made from the conflicting mass of data three considerations should be kept in mind:

(1) Given the condition of relative ignorance regarding the critical parameters for obtaining the phenomenon together with the likelihood that the phenomenon is dependent upon complex interactions among multiple variables, it does not seem unlikely that difficulties will be encountered in replicating procedures and results within a given laboratory as well as between various laboratories.

(2) Under conditions of low replicability within and between laboratories it becomes meaningful to evaluate the phenomenon by reference to the totality of relevant data.

(3) As a consequence of both statistical and methodological considerations a significant positive result must be given more weight than a null result when toting up the subjective probabilities.

One of the problems in such a procedure is the difficulty in being confident that one is in fact dealing with all or nearly all of the experiments which have actually been completed on the problem. In an attempt to solve this problem I have written to 40 researchers who have published in this field asking them to report any completed experiments which they had not published. Twenty of these researchers responded, and on the basis of their replies and other knowledge of unpublished data, I feel quite comfortable in concluding that there is simply no reason whatever to believe that a really substantial number of null result studies have been conducted and not reported either by publication, preprints, reports at conferences, or personal communication. Having been assured on this point I have proceeded to assess the number of studies which support the transfer phenomenon, the number which obtained null results and those which are equivocal. Since many of the papers report several experiments and multiple comparisons within a given experiment I have indicated the number of independent experiments which are relevant to the phenomenon rather than 'counting papers'. Review papers are included only if they include reports of experiments not reported elsewhere. The results of this effort may be seen in table 13.3. It was, of course, occasionally difficult to decide which category a particular experiment should be in since some

TABLE 13.3

Categorization of published and unpublished experiments on memory transfer as supporting the effect (+), null effect (−) or equivocal (0).

Experimental report	+	−	0
Ádám and Faiszt (1967)	7	5	0
Albert (1966)	0	1	0
Allen et al. (1969)	0	1	0
Babich et al. (1965a)	1	0	0
Babich et al. (1965b)	1	0	0
Beatty and Frey (1966)	0	1	0
Bonnett (1967)	1	1	0
Branch and Viney (1966)	0	1	0
Braud (1970)	2	0	0
Byrne et al. (1966)	0	18	0
Byrne and Samuel (1966)	4	0	0
Byrne and Hughes (1967)	1[a]	–	–
Carran and Nutter (1966)	1	0	0
Chapouthier and Ungerer (1969)	1	1	0
Chapouthier et al. (1969)	2	1	0
Corson and Enesco (1968)	0	8	0
Daliers and Rigaux-Motquin (1968)	8	0	0
Daliers and Giurgea (1971)	1	0	0
De Balbian Verster and Tapp (1967)	0	2	0
Dyal and Golub (1967)	0	0	1
Dyal et al. (1967)	2	0	0
Dyal and Golub (1968)	1	1	0
Dyal (1969)	0	1	0
Dyal and Cornell (1969)	1	0	0
Dyal and Golub (1969)	0	0	1
Dyal et al. (1969)	1	1	0
Dyal and Golub (1970a)	0	1	0
Dyal and Golub (1970b)	0	1	0
Essman and Lehrer (1966)	0	0	1
Essman and Lehrer (1967)	1	0	0
Faiszt and Ádám (1968)	4	0	0
Fjerdingstad et al. (1965)	1	0	0
Fjerdingstad (1969a)	1	0	0
Fjerdingstad (1969b)	3	0	3
Fjerdingstad (1969c)	4	0	0
Fjerdingstad et al. (1970)	2	0	0
Gay and Raphelson (1967)	2	0	0
Gibby and Crough (1967)	1	0	0
Gibby et al. (1968)	1	0	0
Golub and McConnell (1968)	1	0	0
Golub et al. (1969)	1	0	0
Golub et al. (1970)	2	0	3
Gordon et al. (1969)	0	1	0
Greene and Kimble (1967)	0	1	0
Gross and Carey (1965)	0	1	0
Gurowitz (1968)	0	2	0
Halas et al. (1966)	0	1	0
Hayes (1966)	0	1	0
Herblin (1970)	3	0	0
Hoffman et al. (1967)	0	1	0
Hutt and Elliott (1970)	0	1	0
Jacobson et al. (1965)	1	0	0
Jacobson et al. (1966a)	1	0	0

Experimental report	+	–	0
Kimble and Kimble (1966)	0	1	0
Kleban et al. (1968)	1	0	0
Krech et al. (1967)	–	–	[e]
Lagerspetz et al. (1968)	0	2	0
Lagerspetz (1969)	1	2	0
Lambert and Saurat (1967)	0	1	0
Luttges et al. (1966)	0	8	0
McConnell et al. (1970)	8	1	1
Malin et al. (1970)	1	0	0
Miller (1967)	0	2	0
Miller et al. (1969)	0	5	0
Moos et al. (1969)	1	0	0
Nissen et al. (1965)	1	1	0
Reiniš (1965)	1	0	0
Reiniš (1968)	1	0	0
Reiniš and Kolousek (1968)	1	0	0
Reiniš (1969a)	1	0	0
Reiniš (1969b)	1	0	0
Reiniš and Mobbs (1970)	1	1	0
Revusky and De Venuto (1967)	0	0	1
Røigaard-Petersen et al. (1968)	3	10	0
Rosenblatt et al. (1966a)	2	0	0
Rosenblatt et al. (1966b, c)	4	1	2
Rosenblatt and Miller (1966a)	3[b]	4	0
Rosenblatt and Miller (1966b)	3	0	0
Rosenblatt (1970)	1[c]	–	–
Rosenthal and Sparber (1968)	1[a]	0	0
Rucker and Halstead (1970)	1	2	1
Smits and Takemori (1968)	0	4	0
Theologus (1967)	0	1	0
Tirri (1967)	0	3	0
Ungar and Oceguera-Navarro (1965)	1	0	0
Ungar (1966)	2	0	0
Ungar and Cohen (1966)	3	1	1
Ungar (1967a)	3	0	0
Ungar (1967b)	2	0	0
Ungar and Irwin (1967)	6[d]	1	0
Ungar et al. (1968)	3[d]	0	0
Ungar (1969)	2	2	0
Ungar and Fjerdingstad (1969)	2	0	0
Ungar and Galvan (1969)	1[a]	0	0
Weiss (1970)	1	0	0
Wolthuis (1970)	4	8	0
Wolthuis et al. (1969)	1[a]	0	0
Zippel and Domagk (1969)	1	0	0
	133	115	15

[a] It is not possible to determine from the available abstract how many separate experiments were run; a conservative categorization is thus to count only one positive.

[b] This set of experiments involved exploration of many dosages, extraction procedures and injection sites. Whenever a particular training procedure was shown to transfer with at least one chemical procedure the experiment was counted as positive.

[c] This paper does not lend itself to classification of the experiments according to the criteria used in this table. In 16 of 18 identical experiments, positive transfer of an asymmetrical discrimination was transferred. It is unclear how many of the individual experiments reached significance, but the overall effect is highly significant. I have thus taken the conservative stance of counting this very substantial paper as a single positive instance.

[d] Several experiments in this report were parametric studies. If a significant effect was obtained for at least one level of the parameter, the experiment was counted as a positive instance.

[e] The number of experiments run is unknown at this point but the overall results are equivocal.

dependent variables indicated a significant effect but others did not. I chose to class the experiment as supporting the validity of the phenomenon if any dependent variable showed a statistically reliable effect at any point in the test series. The reader may regard this as an excessively liberal; however, it seemed especially useful and meaningful in categorizing a series of experiments such as those reported by Rosenblatt and Miller (1966a, b) or by Ungar and Irwin (1967) in which various dosage and temporal parameters were being explored and positive results would not be expected in many of the experiments. Further, it should be noted that a quite conservative stance was taken regarding the p value necessary to reject the null hypothesis. Since there is ample evidence that both significant positive transfer and negative transfer (inversion) effects are obtained and that these may depend on type of task, dosage etc., it is clear that *the use of one-tailed tests at the 0.05 level cannot be justified* as indicating a statistically reliable effect. Therefore in the present table only experiments reporting effects significant at the 0.025 level (one-tail) or 0.05 level (two-tail) are accepted as positive demonstrations of the effect.

It may be seen from table 13.3 that of all the experiments (published and unpublished) which have thus far been reported 133 have yielded significant transfer effects, 115 have yielded null results and 15 are equivocal. Given the three considerations previously noted and especially the point that a positive experiment must be weighted more than a negative experiment the conclusion is inescapable: the memory transfer effect is a real phenomenon!

13.4. The question of specificity

The term 'specificity' occurs quite frequently in the various technical languages which are relevant to memory transfer. Papers presented at the Tihany symposium referred to 'molecular specificity', 'information specificity', 'system specificity' and other unspecified specificities. I refer to 'behavioral specificity' in the present context when I use the term specificity. The problem of behavioral specificity may be stated simply: in what ways and to what extent are the behaviors of the recipients dependent upon the experiences of the donors? The question was originally raised in the context of whether or not learning or memory was being transferred from one organism to another. However, it may be seen that the chemical transfer of such behavioral phenomena as stimulus sensitization and pseudo-conditioning would be as remarkable and important as those behaviors which involve classical or instrumental conditioning. It is apparent that if there were not some relationship between the behavior of the donors and the behavior of the recipients it would not be possible to infer a transfer effect at all. Thus any demonstration of behavioral transfer via injection involves some degree of behavioral specificity. Even so it is legitimate, and even important, to raise questions regarding the type and degree of correspondence between donor and recipient behaviors; i.e. to attempt to characterize the specificities which are involved.

We can explicate the problem of specificity most readily by considering several examples. Let us suppose that experimental donors are trained in a Skinner box to approach a food cup for food whenever a click occurs; the control donors simply 'rest' in their home cages. Now if, as in the Babich et al. (1965b) experiment, the experimental recipients make more magazine-approach responses to a test click than do the control recipients, what type of specificity has been transferred? Clearly an increased tendency to approach the food cup has been transferred and it is dependent upon the differential prior experience of the experimental donors. However, it is unclear which aspect of the experience of the experimental donors the transfer is dependent upon. (a) Is the *contingency* between the click and food reward the critical event? If so, then the transfer would be *transfer of learning* in the strictest sense. (b) Is the effect dependent upon simply having had the sensory stimulation of the click and thus is *transfer of stimulus sensitization*? (c) Is the effect dependent upon simply having been trained in a Skinner box with a resultant tendency to be more reactive to all stimulation, i.e. to be more activated or aroused?

The plausibility of alternative (c) could be diminished if it could be shown that the transfer effect is dependent upon the type of stimulus used in training the donors. Jacobson et al. (1965) trained two groups of rats to approach the food cup. In one group the discriminative stimulus was a distinctive click; for the other group the stimulus was a light. One group of recipients was injected with click-trained brain extract, the other with light-trained brain extract. Both groups were then tested by determining the number of cup responses made to 25 click stimuli and 25 light stimuli presented randomly. The click-injected group responded to the click reliably more than to the light, whereas the light-injected group responded more to the light than to the click.

This experiment shows that the transfer is specific to the stimulus to which the animal has been previously exposed and thus brings into question the validity of a general activation–general sensitization interpretation. However, the experiment does not permit us to determine if the transfer is dependent upon reward following the critical stimulus, i.e. we cannot differentiate between transfer of learning and transfer of sensitization to a specific stimulus. The obvious control for stimulus sensitization was lacking in that the donor groups did not have exposure to both stimuli but with only one or the other reinforced. The critical experiment using differential training within-subjects has not been done. However, the control procedures used in our single bar series are relevant to the question (Dyal et al. 1967; Dyal and Golub 1968). It will be recalled from the previous section that these experiments used a yoked-control group which had non-reinforced experience with the click. Since significant transfer effects were obtained when the experimental group was compared with this sensitization control the plausibility of alternative (b) is brought into question. Although the Dyal et al. experiments cannot completely eliminate alternative (c) (general activation) they provide data which suggest that activation cannot be the whole story (see Dyal and Golub 1968, 1970). Furthermore, Golub and McConnell (1968) have shown that the critical experimental versus control group differences occur only during the first 15 sec of the 60 sec interstimulus interval. If the effect were simply due to differential activation, it would not be expected to be limited to the period shortly after the presentation of the click stimulus.

We may conclude that the most plausible interpretation of the transfer effects reported by Babich et al. (1965b), by Jacobson et al. (1965) and by Dyal et al. (1967) is that the effect represents a specific transfer of learning. However, the argument would be strengthened if significant transfer effects were demonstrated in an experiment in which donor group 1 was differentially conditioned with S^D-light, S^Δ-click; while group 2 had S^D-'click' and S^Δ-'light'. Results, indicating that recipients injected with 'S^D-light, S^Δ-click' make more approach responses to the light while those injected with 'S^D-click, S^Δ-light' made more to the click, would constitute extremely strong evidence for specific transfer of learning. A conceptually similar experiment was conducted by Gurowitz (1968) but no memory transfer was obtained.

Golub and I were quite early persuaded that demonstrations of specific transfer of learning required within-S experimental designs involving discrimination learning or differential conditioning. In 1967 we conducted two experiments of this sort; the first of these emphasized differences in the stimuli between two situations and minimized response differences, while the other minimized stimulus differences and emphasized differences in response morphology.

In the first experiment (Dyal and Golub 1967), the donors learned a black–white discrimination in the Grice-type apparatus used by Ludvigson and Gay (1966). We trained the donors against their original preference 12 trials/day for nine days. All subjects reached a criterion of 18/20 correct choices. The recipients were given 20 non-reinforced preference tests and were then injected with a brain homogenate of 1.5 brain equivalents in 3.0 ml. The recipients were injected with the brain homogenate from animals which had been trained to the brightness opposite to the non-reinforced preinjection preference of the recipients. Twenty-four hrs after injection, while the recipients had been food-deprived for 48–56 hrs, they were given 10 non-reinforced preference trials. The basic datum for each animal was whether his post-injection preference changed in the direction of his injection, showed no change or changed in the opposite direction. Twelve of the 18 recipients changed their preference in a direction consonant with their injection, three changed in the opposite direction and three did not change. Analysis of the resulting choice data using the sign test approximation of the binomial expansion resulted in a highly significant ($p = 0.018$) statistic. However, since the sign test does not make use of scores in which there is no change it can be considered to be a liberal test of the hypothesis and some test which would take these scores into account might be more appropriate. Since the animals which showed change did so in a proportion of 4:1 (12:3) favoring the hypothesis it would seem to be conservative to assign the 'no-change' animals to the positive and negative categories in the proportion of 2:1. Under such a test the resulting statistic is still significant at the 0.015 level. The most conservative approach is to consider that all subjects who did not show change are negative cases for the hypothesis. Under this assumption the experimental hypothesis of transfer of stimulus specific preference must be rejected since the associated p value is 0.13. Considering that the burden of proof must be on those who propose that the transfer is specific we felt that it was appropriate to take the conservative position. This stimulus preference experiment thus must be interpreted as encouraging but not convincing evidence for the hypothesis that transfer of behavioral bias is specific to the training situation of the donors. Results which are less encouraging have been reported by Wagner et al. (1966), Kimble and Kimble (1966), and Greene and Kimble (1966) in attempts to transfer brightness discrimination in T-maze or Y-maze situations. However, as Wagner et al. (1966) point out: 'The conclusions which may be drawn from this failure to observe a transfer effect are obviously limited. The study was not designed as an exact replication of either the Fjerdingstad et al., or Jacobson et al. studies and hence offers no necessary challenge to their reproducibility. There are indeed many parameters of unknown significance to the production of transfer that must be considered'.

Treatment of experimental groups in Dyal and Golub's (1969) response specificity design.

	Donor training	Injection	Recipient tests	
			Behavioral bias[a]	Learning enhancement[b]
Group BP	$S_B \to R_{BP} \to +$ $S_M \to R_{ME} \to -$	$R_{BP}^+R_{ME}^-$	R_{BP} or R_{ME}	$R_{BP}^+R_{ME}^-$ $R_{BP}^-R_{ME}^+$
Group ME	$S_B \to R_{BP} \to -$ $S_M \to R_{ME} \to +$	$R_{BP}^-R_{ME}^+$	R_{BP} or R_{ME}	$R_{ME}^+R_{BP}^-$ $R_{ME}^-R_{BP}^+$

[a] Non reinforced tests
[b] One response reinforced

Fig. 13.3. Experimental paradigm for Dyal and Golub's response transfer experiment.

In the 'response-transfer' experiment from our laboratory, Golub and I trained two groups to make responses which were morphologically quite distinct from each other. The basic experimental paradigm is represented in fig. 13.3. Group BP was rewarded for bar pressing but not for magazine entry (ME). Group ME was rewarded for magazine entry but not for bar pressing. The magazine-entry response involved sticking the nose into the cup in which food is normally delivered in the Grason–Staedler Skinner box; but note that in this case, the food pellets were delivered into a cup in another part of the box. For the donors a bar press (magazine entry) resulted in the delivery of a food pellet and a 'light–click' on a CRF schedule. Following the completion of training (18 days, 30 min per day) the donors were sacrificed and a brain homogenate prepared using the same procedures as those described by Dyal et al. (1967). Half of the recipients were injected with homogenate from donors trained $BP(^+)ME(^-)$ and the other half from donors trained $ME(^+)BP(^-)$. Twenty-four hours after injection they were given a non-reinforced test of behavioral bias; the 'light–click' was presented every

60 sec for 30 min. The number of ME and BP made during this period was counted. If the transfer effect is response-specific then those recipients receiving homogenate from BP-trained donors should increase their bar-pressing more than those injected with ME-trained homogenates. Similarly those recipients injected with ME-trained homogenate should increase their ME output more than the BP-injected subjects. The results only partially confirm the hypothesis of specific transfer of response bias. The bar press data supported the hypothesis in that BP-injected recipients significantly increased the bar presses over baseline whereas the ME-injected recipients showed a slight decline ($p = 0.045$; table 13.4). However, argument for specific transfer of response bias is weakened by the fact that the BP-injected recipients also tended to make more ME responses than did the ME-injected recipients ($p < 0.10$, two-tailed test). The ME-injected recipients showed no tendency to increase their ME, in fact, as may be seen in table 13.5, the post-injection score is somewhat lower than the preinjection score.

TABLE 13.4

Mean number of bar presses during 30 min preference test.

Injection received			
BP		ME	
pre	post	pre	post
7.45	10.15	8.87	6.19

TABLE 13.5

Mean number of magazine entries during 30 min preference test.

Injection received			
BP		ME	
pre	post	pre	post
57.8	58.5	55.4	46.2

The test of learning enhancement consisted of five, daily, 30 min sessions in which half of each injection group was reinforced for either BP or ME. If there is a response-specific enhancement of learning then those recipients which are trained to the same response as their injection (BP–BP and ME–ME) should show greater enhancement than those which are trained to make a response which conflicts with their injection bias (BP–ME and ME–BP). The only test session in which differences appeared between the groups was the first reinforced session. This session was held immediately following the non-reinforced preference test previously discussed. Considering first those groups which were trained to bar press we find that the BP-injected recipients (group BP–BP) made significantly more bar presses than those which were injected with ME-trained homogenate (group ME–BP). Taken with the similar preference test data, these results are consistent with the assertion that the transfer effect is specific to the particular response which has been trained. However, this conclusion is again very much weakened by the fact that the BP–ME groups made more magazine entries than the ME–ME.

The implication is, of course, that since both of the BP-injected groups were more responsive than the ME-injected animals on the operant for which they were reinforced, the total effect could be explained on the basis of the BP-injected animals being more generally active or aroused than the ME-injected animals. Furthermore, the preference test data is consistent with this interpretation since the BP-injected group made both more BP responses and more ME responses than the ME-injected group.

On the other hand, Golub and I have argued elsewhere that there are several aspects of the data which are not consistent with a general activation interpretation (Dyal and Golub 1970). Nonetheless, it would be fair to say that the results of these two studies from our laboratory do not strongly confirm the hypothesis that the transfer effects are specific to either the critical stimuli or the type of response which is reinforced during training.

What light is shed on the problem of specificity by research from other laboratories? The first experiment dealing with the problem was that of Jacobson et al. (1965). It will be recalled that recipients injected with light-trained brains responded to the light more than the click, whereas the opposite was true for click-injected animals. As was previously noted, this experiment permits us to conclude that the memory transfer was specific to a stimulus which was present in the donor's training, but because of the lack of sensitization controls it does not permit us to state that specific transfer of *learning* is involved. The two attempts which have been made to replicate this experiment have been unsuccessful (Halas et al. 1966; De Balbian Verster and Tapp 1967). It seems clear that this experiment should be repeated with sensitization control groups.

Ungar (1966) has reported what he calls a 'cross-transfer' design from which he infers that transfer of habituation of a startle response is specific to the stimulus used to habituate the donors. The startle response in one group of mice was habituated to an air-puff; a second group was habituated to a loud noise. The recipients were injected with either sound-habituated extract or puff-habituated extract. Half of each of these injection groups was then habituated to either the puff or the noise. The habituation of the recipients was greatly facilitated when their injection was from donors who had previously been habituated to the same stimulus. It was facilitated much less, if at all, by injection of extracts from donors habituated to the alternate stimulus. These results are consistent with the notion that the transfer effect may be specific to the stimulus modality which was used in the original training. Unfortunately, this important experiment has not been replicated.

Ungar has also used the cross-transfer procedure to demonstrate that avoidance in a step-down task is not facilitated by injections of 'dark-box avoidance extract'; similarly dark-box avoidance is not facilitated by injection of 'step-down avoidance extract'. On the other hand, highly significant passive avoidance is obtained when recipients are tested in situations corresponding to their injections. Both of these cross-transfer experiments do, indeed, permit us to infer that the transfer is at least specific to the gross characteristics of the original training situations. However, a previous argument seems relevant to this point.

'In any learning situation there is an extensive set of stimulus elements which are conditioned to an equally extensive set of response elements. If no transfer is observed when recipients are tested in a stimulus situation which is quite unlike the donor's training situation and which requires a different kind of response, the most one can conclude is that there is at least one behavior in one situation which is not activated by the transfer agent. Hardly an earth shaking inference! In other words, the strength of the argument for specificity of transfer increases as the number of stimulus and response elements increase between the training and the test situation' (Dyal 1971). Thus the strongest test of specificity would be within a single stimulus or response dimension in which case the test would be based on differences in slope of generalization gradients between experimental and control groups. In the case of the passive avoidance response to a black box one procedure would be to vary the brightness of the dark side from black to light grey using the standard procedures to test for primary stimulus generalization. However, it should be noted that Ungar's cross-transfer groups at least serve the function of a quasi foot-shock control group. This permits us to infer that the transfer of avoidance in the step-down box is not simply based on some generalized increase in fear based on having been shocked. The experiment does not permit us to infer that the transfer is based on transfer of fears conditioned to specific stimuli or of particular types of response conditioned in the donors.

The reason for the inadequacy is that the test stimuli/responses for the cross-transfer animals are different from those of the regular transfer animals (and the donors) along many complex dimensions. We thus cannot differentiate between the following inferences: (a) the transfer is based on the instrumental conditioning of passive avoidance in the donors; (b) the transfer is based on the classical conditioning of fear to the stimuli of the box; (c) the transfer is based on instrumental passive avoidance conditioning mediated by the classical conditioning of fear. Despite the fact that Ungar's cross-transfer of passive avoidance experiment cannot differentiate these alternatives it represents a potentially important approach and deserves to be repeated with standard foot-shock control groups accompanying each of the donor training groups.

Rosenblatt and his colleagues have tended to use a variety of discrimination tasks in which the basic comparison in the test situation is the proportion of left-responses made by the left-injected group with the proportion of left-responses made by the right-injected group. Rosenblatt and Miller (1966a) reported the results of seven different training procedures in which a search had been made for a combination of dosage, injection procedure and extraction method which would yield significant transfer. In three of the seven experiments at least one procedure yielded significant transfer. In this series and in the follow up series using symmetrical choice tasks (Rosenblatt and Miller 1966b) both significant positive transfer and significant inversion effects were obtained. The direction of the effect seemed to be a complex function of task characteristics and dosage. In more recent work Rosenblatt (1969) reported a series of 15 experiments in a symmetrical two-bar box situation. The results confirmed his early work in suggesting a peak of positive transfer at 0.025 brain equivalents and 'inversion effects' (negative transfer) being obtained at 0.013 and 0.050 brain equivalents. In addition, Herblin (1969) using the same procedures and equipment, replicated the positive peak at 0.025 brain equivalents. Nonetheless, as Rosenblatt continued to repeat the experiment he found that '...the predicted maximum, which had never been completely reliable, could no longer be obtained at the same dose; in fact systematic inversion effects began to appear at 0.025 brains' (Rosenblatt 1969). As a result of his failure to obtain a reliable positive transfer with a specific dosage in a symmetrical task Rosenblatt decided to move to an asymmetrical test situation in which both response requirements and stimulus associated with two alternatives were quite distinctive. In this series of 18 experiments, 16 resulted in positive transfer and only two in slight, non-significant negative transfer. The probability of the combined effect was less than 0.0002. Since the discrimination procedure seems so obviously to involve transfer of training relevant information it seems reasonable to agree with Rosenblatt: 'The fact that systematic differences in left or right bias can be induced, however, suggests that a training specific mechanism, rather than a general polarizing effect (such as induction of fear or curiosity) is involved in these experiments' (Rosenblatt 1969). Despite the obvious 'face validity' of these results we should not forget that it has not been demonstrated that the behavior induction is specific to the presence of response-contingent reward in a particular environment, i.e. to the transfer of instrumental conditioning. On the contrary it seems likely that the effect is based on a tendency to approach and spend time in a particular stimulus situation. It is the learning of a positive stimulus valence associated with reward rather than learning to make a particular response. Rosenblatt comments that '...present evidence suggests that differences in response to stimuli can be induced more readily and more reliably than differences in operant behavior' (Rosenblatt 1969, page 668). This view is consistent with the fact that in the Dyal et al. (1969) experiment the Y^+ group tended to perform like the other rewarded groups when magazine entry was the dependent variable (thus presumably responding to the positive valence of the 'light–click' stimulus) but like the non-rewarded groups when bar press was the dependent variable (see tables 13.1 and 13.2).

Although the Rosenblatt experiments with the asymmetrical door/bar task strongly suggest 'training-specific transfer' effects, two reservations should be noted: (1) the experiments do not permit us to infer what aspect of the donor's experience is critical for the transfer. Is the fact that reward was contingent on differential response critical? Is the effect based on the conditioning of a positive valence for stimuli associated with reward (secondary reinforcement)? Is the effect based on the greater duration of exposure to the stimuli associated with reward, i.e. stimulus sensitization? The appropriate control procedures necessary to differentiate these alternatives are straightforward. Although it may be premature to ask that these controls be run in future experiments, the nature and degree of specificity will not be discernible without them. (2) The absolute size of the effect is very small and requires a large number of experiments with a large number of recipients in each to be detected.

What may we conclude with regard to the behavioral specificity of the transfer effect?

(1) Discrimination studies such as those conducted by Rosenblatt and by Dyal and Golub represent the strongest paradigm for demonstrating the effect; and Rosenblatt's (1970) results in the asymmetrical door/bar task are strongly suggestive that indeed the transfer may be relatively specific to the total task of the donors.

(2) The cross-transfer paradigm proposed by Ungar suffers certain weaknesses but can be strengthened by attention to the specific dimensions which may be critical for the transfer.

(3) Both the discrimination paradigm and the cross-transfer paradigms must be repeated with appropriate controls for sensitization and 'motor-training' (see Essman and Lehrer 1967) before it can be convincingly argued that the transfer effect is specific to the training of the donors and does indeed reflect a 'reinstatement' of the learning process which occurred in the donors.

13.5. Directions for future research

It seems quite reasonable to think that future research on the problem of memory transfer will be directed toward the following questions among others.

13.5.1. Robust and reliable behavioral assays

Is it possible to discover a behavioral assay which will provide robust effects which are replicable between laboratories? It has taken five years to establish that indeed the memory transfer effect is real. Furthermore, its 'reality' is based upon the 'overwhelming preponderance of the data' kind of argument rather than upon the stronger base of a highly reliable technique for producing the effect. It is quite clear that first priority must still be given to discovering a behavioral assay technique (and extraction procedure) which has high replicability across many laboratories. However, the first step in achieving inter-laboratory replicability is the demonstration of a replicable and robust effect within a given laboratory in experiment after experiment. At this point Rosenblatt's asymmetrical door/bar experiments seem to provide the best evidence for a consistently replicable effect within a given laboratory. Unfortunately, the effect does not appear to be as robust within a given experiment as would be desirable. All of the other major paradigms such as the Gay and Raphelson/Ungar dark-box passive avoidance, Rosenblatt's symmetrical double bar procedure, and Dyal and Golub's AER procedure can claim a modest amount of both intralaboratory *and* interlaboratory replicability. They thus offer promising leads for further research.

13.5.2. Relevance to learning

Is the transfer effect relevant to learning processes in the donors? There are two aspects to this issue. One is the necessity for a better controlled and more incisive demonstration of behavioral specificity. This problem has previously been discussed at length. The second aspect is the necessity to examine the transfer of other major phenomena of learning other than simple response acquisition or discrimination. The case for the relevance of memory transfer to the problem of the molecular basis of learning would be greatly enhanced if it could be demonstrated that partial reinforcement extinction effects, primary stimulus generalization effects, regular extinction effects, latent extinction effects, reward magnitude effects, etc., etc., could be transferred. Thus far, the relevant evidence is minimal; Golub et al. (1970) and Braud (1970) present data which suggest that extinction processes may be transferable while Dyal and Golub (1970) were unsuccessful in transferring a partial reinforcement extinction effect.

13.5.3. Disruptors of memory consolidation

What are the effects of disruptors of memory consolidation on memory transfer? It seems likely that future research will make use of various techniques which have been used to disrupt memory consolidation. One such approach would be to determine if the transfer of the passive avoidance (e.g. in the step-down apparatus) would be blocked by electroconvulsive shock delivered at an appropriate time interval following the conditioning trial. A similar strategy has been followed by Reiniš and his collaborators in their attempts to determine the effects of various anti-metabolites on memory transfer (see chapter 7 in this volume).

13.5.4. Modifiability of genetically controlled behavior

Can genetically controlled behaviors be modified by transfer techniques? Demonstration of a positive answer to this question could only be regarded as revolutionary. Three experiments have been reported in which an attempt has been made to modify, by injection of 'antagonistic' brain material, a behavior pattern which is believed to be under genetic control. Golub and I (1969) noted that alcohol preference is a polygenically controlled behavior in mice. It was conjectured that level of alcohol preference was controlled by extent of repression of these genes which might be derepressed by injections of brain extracts from a strain which normally had a different preference level. Fifteen mice of the C57 strain and 15 of the DBA strain were given continuous access to two solutions, tap water and 10% ethanol. After four days of preference testing the preferences of the two groups were clearly established; alcohol was 95% of the daily liquid intake for the C57 strain and only 5% of the intake for the DBA strain.

The two groups were divided at random into donors (n = 14) and recipients (n = 16). The donors were decapitated, their brains removed and frozen immediately on dry ice. RNA extractions were by the phenol method. The RNA was injected intraventricularly into recipients in the amount of 0.45 ml (0.25 brain equivalents). Four DBA were injected with C57 extract and four with DBA extract; similarly four C57 were injected with C57 extract and four with DBA extract. Liquid intake was measured every 4 hrs for two days following injection. All recipients showed significant suppression of fluid intake during the first 24 hrs following injection. This was undoubtedly due to a general systemic effect of the injection. There were no differential effects of brain RNA injections on either the total liquid intake for any of the strain-injection combinations. These results are consistent with the reported failure by Reiniš and Mobbs (1969) to transform 'killer' rats into 'non-killer' rats by injections of brain homogenates from 'non-killers'. Similarly Lagerspetz (1968) failed to transfer differential activity level from 'high-active' and 'low-active' mice to mice of an intermediate activity level. On the other hand, Coward has obtained suggestive evidence that substrate preference can be modified by injections of RNA from the hepato-pancreas of two species of crabs (*U. pugilator* which tends to burrow in sand and *U. pugnax*, which burrows less often and tends to prefer mud to sand; McConnell and Shelby 1970, pages 99–100).

It is of course premature to conclude that genetically controlled behavior cannot be modified via transfer procedures but it does seem certain that further attempts must be much more sophisticated and probably will require intervention during the critical gestation periods when the fetal brain is being formed. Whether or not a positive answer is eventually forthcoming to this question, it seems that the question of interaction between genetic and training variables noted by Carran and Nutter (1966) should be pursued further. They trained donor mice of two different strains to approach a food cup when a click was sounded. They found an interaction between the genetic variable (strain) and direction of transfer such that '...recipients of same-strain trained or different-strain untrained approached the cup faster than those receiving either same-strain untrained or different-strain trained extract (Carran and Nutter 1966).

13.5.5. Comparative analysis

Are there other species which offer special advantages in the study of the biochemistry of learning via memory transfer? Perhaps the cleanest and most persuasive memory transfer study that has thus far been conducted is that of Jacobson et al. (1966b). In the rush from planaria to mammals this study has been neglected. It demonstrated in two experiments robust transfer of classically conditioned contractions in planaria; furthermore, it showed that extinction did *not* transfer. This study is important and needs replicating. Furthermore, the classical conditioning paradigm offers the possibility for determining the degree of stimulus specificity through the use of standard primary stimulus generalization tests. At this time transfer has been demonstrated in planaria, rats, hamsters, mice, chicks (Rosenthal and Sparber 1968, chapter 9 this volume) and goldfish (Zippel and Domagk 1969; see chapter 11 in this volume). There seems to be little point in simply multiplying the number of species in which the effect may be demonstrated; rather, preparations should be selected because they offer unique advantages to solving particular problems. The rat is an ideal subject because so much is known at the behavioral level about learning processes in this animal; the mouse offers advantages in studying genetic–environment interactions; chicks may be especially useful because of the early lack of blood–brain barrier; comparison of goldfish and pigeons in comparable transfer tasks may be useful in further elucidating Bitterman's contention that learning in the two species may involve fundamentally different mechanisms (Behrend et al. 1970). Other preparations offer the possibility of innovative approaches, for example, Pietsch and Schneider (1969) have attempted to obtain behavioral induction via whole-brain transplants in salamanders. It would appear that annelids would also provide a unique opportunity for whole-brain transplants, both in the coelom and in the supra-esophageal ganglion cavity.

13.5.6. Chemical nature of the active agent

What is the chemical nature of the active transfer agent? Two major points of view have emerged thus far. One suggests that the critical coding takes place at the level of RNA (see Fjerdingstad et al. 1965; Jacobson et al. 1966a,b; McConnell et al. 1967; Faiszt and Ádám 1968). The other approach suggests that peptides are critical (see Rosenblatt et al. 1966c; Ungar and Irwin 1967). More recent work by Ungar and Fjerdingstad (1969) provides a compromise view which emphasizes RNA-bound peptides. Golub et al. (1969) have found that peptide containing preparations, RNA extracts and homogenates were all effective in mediating the transfer of a passive avoidance response. Further efforts in more precise determination of the nature of the active fractions (which may be dependent on the particular task) will undoubtedly be forthcoming.

References

ÁDÁM, G. and J. FAISZT, 1967, Nature *216*, 198.

ALBERT, D. J., 1966, Neuropsychologia *4*, 79.

ALLEN, A. R., R. J. GRISSOM and C. L. WILSON, 1969, Psychon. Sci. *15*, 257.

APPLEWHITE, P. and F. T. GARDNER, 1968, Nature *220*, 1136.

BABICH, F. R., A. L. JACOBSON and S. BUBASH, 1965a, Proc. Natl. Acad. Sci. U.S.A. *54*, 1299.

BABICH, F. R., A. L. JACOBSON, S. BUBASH and A. JACOBSON, 1965b, Science *149*, 656.

BARONDES, S. H. and S. D. COHEN, 1967, Brain Res. *4*, 44.

BEATTY, W. W. and P. W. FREY, 1966, unpublished paper.

BEHREND, E. R., A. S. POWERS and M. E. BITTERMAN, 1970, Science *167*, 389.

BONNETT, K. A., 1967, Behavioral specificity in chemical interanimal transfer of training. Paper presented at the Western Psychological Association Meetings, San Francisco, April 25.

BRANCH, J. C. and W. VINEY, 1966, Psychol. Rep. *79*, 923.

BRAUD, W. G., 1970, Science *168*, 1234.

BYRNE, W. L. and A. HUGHES, 1967, Fed. Proc. *26*, 676.

BYRNE, W. L. and D. SAMUEL, 1966, Science *154*, 418.

BYRNE, W. L., D. SAMUEL, E. L. BENNETT, M. R. ROSENZWEIG, E. WASSERMAN, A. R. WAGNER, R. GARDNER, R. GALAMBOS, B. D. BERGER, D. L. MARGULES, R. L. FENICHEL, L. STEIN, J. A. CORSON, H. E. ENESCO, S. L. CHOROVER, E. C. HOLT, III, P. H. SCHILLER, L. CHIAPPETTA, M. E. JARVIK, R. C. LEAF, J. D. DUTCHER, Z. P. HOROVITZ and P. L. CARLSON, 1966, Science *153*, 658.

CAMPBELL, B. and J. JAYNES, 1966, Psychol. Rev. *73*, 478.

CARRAN, A. B. and C. B. NUTTER, 1966, Psychon. Sci. *5*, 3.

CHAPOUTHIER, G., B. POLLAUD and A. UNGERER, 1969, Rev. Comp. Animal *3*, 55.

CHAPOUTHIER, G. and A. UNGERER, 1969, Rev. Comp. Animal *3*, 64.

COHEN, H. D. and S. H. BARONDES, 1966, J. Neurochem. *13*, 207.

COOK, L., A. DAVIDSON, D. DAVIS, H. GREEN and E. J. FELLOWS, 1963, Science *141*, 268.

CORNING, W. C. and S. FREED, 1968, Nature *219*, 1227.

CORNING, W. C. and E. R. JOHN, 1961, Science *134*, 1363.

CORSON, J. A. and H. E. ENESCO, 1966, unpublished paper.

CORSON, J. A. and H. E. ENESCO, 1968, J. Biol. Psychol. *10*, 10.

DALIERS, J., 1968, Transfer by a brain extract of a facilitating effect on the habituation of the rat to an acoustic epileptogenic stimulus. Paper presented at 1st International Congress CINS, Milan, October 19.

DALIERS, J. and C. GIURGEA, 1971, Effect of brain extracts on the fixation of experience in rats. *In:* G. Ádám, ed., Biology of memory, Akademiai Kiado, Budapest, in press.

DALIERS, J. and M. I. RIGAUX-MOTQUIN, 1968, Arch. int. Pharmacodyn. *176*, 461.

DE BALBIAN VERSTER, F. and J. T. TAPP, 1967, Psychol. Rep. *21*, 9.

DINGMAN, W. and M. B. SPORN, 1961, J. Psychiat. Res. *1*, 1.

DYAL, J. A., 1969, Incubation of learning and transfer effects. Unpublished research. Abstract distributed at Symposium on Biology of Memory, Hungarian Academy of Science, Tihany, Hungary, Sept. 2.

DYAL, J. A., 1971, Transfer of behavioral bias and learning enhancement: a critique of specificity experiments. *In:* G. Ádám, ed., Biology of memory, Akademiai Kiado, Budapest, in press.

DYAL, J. A. and J. M. CORNELL, 1969, Transfer of conditioned fear via injections of brain homogenates. Unpublished research. Abstract distributed at Symposium on Biology of Memory, Hungarian Academy of Science, Tihany, Hungary, Sept. 2.

DYAL, J. A. and A. M. GOLUB, 1967, J. Biol. Psychol. *9*, 29.

DYAL, J. A. and A. M. GOLUB, 1968, Psychon. Sci. *18*, 13.

DYAL, J. A. and A. M. GOLUB, 1969, Attempt to modify a genetically controlled response via injections of brain RNA. Unpublished research. Abstract distributed at Symposium on Biology of Memory, Hungarian Academy of Science, Tihany, Hungary, Sept. 2.

DYAL, J. A. and A. M. GOLUB, 1970 a, Behavioral transfer via injection of brain homogenate: activation or specificity? *In:* W. L. Byrne, ed., Molecular approaches to learning and memory. Academic Press, New York, pp. 275–284.

DYAL, J. A. and A. M. GOLUB, 1970 b, unpublished research.

DYAL, J. A., A. M. GOLUB and R. L. MARRONE, 1967, Nature *214*, 720.

DYAL, J. A., A. M. GOLUB and H. W. LUDVIGSON, 1969, Further studies of memory transfer, Unpublished paper presented at Meetings of Southwestern Psychological Association. Austin, Texas, April 15 and Western Psychological Association, Vancouver, Canada, May 4.

ESSMAN, W. B. and G. M. LEHRER, 1966, Fed. Proc. *25*, 208.

ESSMAN, W. B. and G. M. LEHRER, 1967, Fed. Proc. *26*, 263.

FAISZT, J. and G. ÁDÁM, 1968, Nature *220*, 367.

FJERDINGSTAD, E. J., 1969a, Nature *222*, 1079.

FJERDINGSTAD, E. J., 1969b, Scand. J. Psychol. *10*, 220.

FJERDINGSTAD, E. J., 1969c, Personal communication.

FJERDINGSTAD, E. J., W. L. BYRNE, T. NISSEN and H. H. RØIGAARD-PETERSEN, 1970, A comparison of 'transfer' results obtained with two different types of extraction and injection procedures using identical behavioral techniques. *In:* W. L. Byrne, ed., Molecular approaches to learning and memory. Academic Press, New York, pp. 151–170.

FJERDINGSTAD, E. J., T. NISSEN and H. H. RØIGAARD-PETERSEN, 1965, Scand. J. Psychol. *6*, 1.

FLEXNER, L. B., J. B. FLEXNER and R. B. ROBERTS, 1966, Proc. Natl. Acad. Sci. U.S.A. *56*, 730.

GAITO, J., J. H. DAVISON and J. MOTTIN, 1968, Psychon. Sci. *13*, 259.

GARCIA, J., D. J. KIMELDORF and R. A. KOELLING, 1955, Science *122*, 157.

GAY, R. and A. RAPHELSON, 1967, Psychon. Sci. *8*, 369.

GIBBY, R. G. and D. C. CROUGH, 1967, Psychon. Sci. *9*, 413.

GIBBY, R. G., D. C. CROUGH and S. J. THIOS, 1968, Psychon. Sci. *12*, 295.

GOLUB, A. M. and J. V. MCCONNELL, 1968, Psychon. Sci. *11*, 1.

GOLUB, A. M. and J. V. MCCONNELL, 1969, unpublished paper.

GOLUB, A. M., L. EPSTEIN and J. V. MCCONNELL, 1969, J. Biol. Psychol. *11*, 44.

GOLUB, A. M., F. R. MASIARZ, T. VILLARS and J. V. MCCONNELL, 1970, Science *168*, 392.

GORDON, M. W., G. G. DEANIN, H. L. LEONHARDT and R. H. GWYNN, 1966, Am. J. Psychiat. *122*, 1174.

GREENE, E. G. and D. P. KIMBLE, 1966, unpublished paper.

GROSS, C. G. and F. M. CAREY, 1965, Science *150*, 1749.

GUROWITZ, E. M., 1968, Psychol. Rep. *23*, 899.

HALAS, E. S., K. BRADFIELD, M. E. SANDLIE, F. THEYE and J. BEARDSLEY, 1966, Physiol. Behav. *1*, 281.

HAYES, J. L., 1966, unpublished paper.

HENDRICKSON, C. W. and R. J. KIMBLE, 1968, Psychon. Sci. 13, 149.

HERBLIN, W. F., 1970, Physiol. Behav., in press.

HOFFMAN, R. F., C. N. STEWART and H. N. BHOGAVAN, 1967, Psychon. Sci. *9*, 151.

HUTT, L. D. and L. ELLIOTT, 1970, Psychon. Sci. *18*, 28.

HYDÉN, H. and P. W. LANGE, 1965, Proc. Natl. Acad. Sci. U.S.A. *53*, 946.

JACOBSON, A. L., F. R. BABICH, S. BUBASH and A. JACOBSON, 1965, Science *150*, 636.

JACOBSON, A. L., F. R. BABICH, S. BUBASH and C. GOREN, 1966a, Psychon. Sci. *4*, 3.

JACOBSON, A. L., C. FRIED and S. D. HOROWITZ, 1966b, Nature *209*, 599.

KATZ, J. J. and W. C. HALSTEAD, 1950, Comp. Psychol. Monogr. *20*, 1.

KIMBLE, R. J. and D. P. KIMBLE, 1966, Worm Runner's Digest *8*, 32.

KLEBAN, M. H., H. ALTSCHULER, M. P. LAWTON, J. L. PARRIS and C. A. LORDE, 1968, Psychol. Rep. *23*, 51.

KRECH, D., E. L. BENNETT and P. RAGAN, 1967, Effects of brain homogenate on reinstatement of early memory. Unpublished paper presented in AAAS symposium, molecular approaches to learning and memory. New York, December 29.

KRECH, D., M. R. ROSENZWEIG and E. L. BENNETT, 1962, J. Comp. physiol. Psychol. *55*, 801.

KRYLOV, O. A., R. A. DANLOVA and V. S. TONGUR, 1964, Life Sci. *4*, 1313.

LAGERSPETZ, K. Y. H., 1969, unpublished paper.

LAGERSPETZ, K. M. J., P. RAITIS, R. TIRRI and K. Y. H. LAGERSPETZ, 1968, Scand. J. Psychol. *9*, 225.

LAMBERT, R. and M. SAURAT, 1967, Bulletin de CERP *16*, 435.

LASHLEY, K. S., 1950, Symp. Soc. Exp. Biol. *4*, 454.

LEAF, R. C., J. D. DUTCHER, Z. P. HOROVITZ and P. L. CARLTON, 1966, unpublished paper.

LEAF, R. C. and S. A. MULLER, 1965, Psychol. Rep. *17*, 211.

LUDVIGSON, H. W. and S. E. GAY, 1966, Psychon. Sci. *5*, 289.

LUTTGES, M., T. JOHNSON, C. BUCK, J. HOLLAND and J. MCGAUGH, 1966, Science *151*, 834.

MACHLUS, B. and J. GAITO, 1968, Psychon. Sci. *12*, 111.

MALIN, D. H., A. M. GOLUB and J. V. MCCONNELL, 1971, Nature, in press.

MCCONNELL, J. V. ,1962, J. Neuropsychiat. *3*, 542.

MCCONNELL, J. V., A. L. JACOBSON and D. P. KIMBLE, 1959, J. Comp. physiol. Psychol. *52*, 1.

MCCONNELL, J. V., T. SHIGEHISA and H. SALIVE, 1970, Attempts to transfer approach and avoidance response by RNA injections in rats. *In:* W. L. Byrne, ed., Molecular approaches to learning and memory. Academic Press, New York, pp. 245–270.

MCCONNELL, J. V. and J. M. SHELBY, 1970, Memory transfer experiments in invertebrates. *In:* G. Ungar, ed., Molecular mechanisms in memory and learning. Plenum Press, New York, pp. 71–101.

MCGAUGH, J., 1967, Proc. Amer. Philos. Soc. *111*, 347.

MILLER, R. E., 1967, personal communication.

MILLER, R. E., W. F. CAUL and I. A. MIRSKY, 1969, unpublished paper.

MOOS, W. S., H. LEVAN, B. T. MASON, H. C. MASON and D. L. HEBRON, 1969, Experientia *15*, 1215.

MURPHY, J. V. and R. E. MILLER, 1955, J. comp. physiol. Psychol. *48*, 47.

NISSEN, T., H. H. RØIGAARD-PETERSEN and E. J. FJERDINGSTAD, 1965, Scand. J. Psychol. *6*, 265.

PALFAI, T. and J. M. CORNELL, 1968, J. Comp. physiol. Psychol. *66*, 584.

PIETSCH, P. and C. W. SCHNEIDER, 1969, Brain Research *14*, 707.

REINIŠ, S., 1965, Activ. Nerv. Super. *7*, 167.

REINIŠ, S., 1968, Nature *220*, 177.

REINIŠ, S., 1969a, Psychon. Sci. *14*, 44.

REINIŠ, S., 1969b, Paper submitted to Activitas Nervosa Superior.

REINIŠ, S. and J. KOLOUSEK, 1968, Nature *217*, 680.

REINIŠ, S. and D. R. MOBBS, 1969, Some applications of 'memory transfer' in the study of learning. *In:* W. L. Byrne, ed., Molecular approaches to learning and memory. Academic Press, New York, pp. 189–193.

REVUSKY, S. H. and F. DEVENUTO, 1967, J. Biol. Psychol. *9*, 18.

RØIGAARD-PETERSEN, H. H., T. NISSEN and E. J. FJERDINGSTAD, 1968, Scand. J. Psychol. *9*, 1.

ROSENBLATT, F., 1969, Proc. Natl. Acad. Sci. U.S.A. *64*, 661.

ROSENBLATT, F., 1970, Induction of behavior by mammalian brain extracts. *In:* G. Ungar, ed., Molecular mechanisms in memory and learning. Plenum Press, New York, pp. 103–147.

ROSENBLATT, F., 1970, Induction of discriminatory behavior by means of brain extracts. *In:* W. L. Byrne, ed., Molecular approaches to learning and memory. Academic Press, New York, pp. 195–242.

ROSENBLATT, F., 1971, Behavior induction of brain extracts: a comparison of two procedures. *In:* G. Ádám, ed., Biology of memory, Akademiai Kiado, Budapest, in press.

ROSENBLATT, F., J. T. FARROW and W. F. HERBLIN, 1966a, Nature *209*, 46.

ROSENBLATT, F., J. T. FARROW and S. RHINE, 1966b, I. Proc. Natl. Acad. Sci. U.S.A. *55*, 548.

ROSENBLATT, F., J. T. FARROW and S. RHINE, 1966c, II. Proc. Natl. Acad. Sci. U.S.A. *55*, 787.

ROSENBLATT, F. and R. G. MILLER, 1966a, Proc. Natl. Acad. Sci. U.S.A. 1423.

ROSENBLATT, F. and R. G. MILLER, 1966b, Proc. Natl. Acad. Sci. U.S.A. 1683.

ROSENTHAL, E. and S. B. SPARBER, 1968, The Pharmacologist *10*, 168.

RUCKER, W. B. and W. C. HALSTEAD, 1969, Memory: antagonistic transfer effects. *In:* W. L. Byrne, ed., Molecular approaches to learning and memory. Academic press, New York, pp. 295–306.

SCHILLACE, R. J., 1968, Diss. Abst. *28*, 3503-B.

SMITS, S. E. and A. E. TAKEMORI, 1968, Proc. Soc. Exp. Biol. Med. *127*, 1167.

THEOLOGUS, G. C., 1967, Thesis, The Catholic University of America, 67–14322.

TIRRI, R., 1967, Experientia *4*, 278.

UNGAR, G., 1965, Fed. Proc. *24*, 548.

UNGAR, G., 1966, Fed. Proc. *25*, 207.

UNGAR, G., 1967a, Fed. Proc. *26*, 263.

UNGAR, G., 1967b, J. Biol. Psychol. *12*, 27.

UNGAR, G., 1968, Memory transfer–specificity and reproducibility. Paper presented at AAAS Symposium, molecular approaches to the central nervous system, Dallas, December 29.

UNGAR, G., 1969, Fed. Proc. *23*, 647.

UNGAR, G. and M. COHEN, 1966, Int. J. Neuropharmacol. *5*, 183.

UNGAR, G. and L. GALVAN, 1969, Proc. Soc. Exp. Biol. Med. *130*, 287.

UNGAR, G. and L. N. IRWIN, 1967, Nature *214*, 453.

UNGAR, G. and E. J. FJERDINGSTAD, 1969, Mol. Neurobiol. Bull. *2*, 9.

UNGAR, G. and C. OCEGUERA-NAVARRO, 1965, Nature *207*, 301.

UNGAR, G., L. GALVAN and R. H. CLARK, 1968, Nature *217*, 1259.

WAGNER, A. R., F. GARDNER and R. GALAMBOS, 1966, unpublished paper.

WEISS, K. P., 1970, Measurement of the effects of brain extract on interorganism information transfer. *In:* W. L. Byrne, ed., Molecular approaches to learning and memory. Academic Press, New York, pp. 325–334.

WINER, B. J., 1962, McGraw-Hill, New York, p. 43.

WOLTHUIS, O. L., 1970, Interanimal information transfer by brain extracts studied in various tests. *In:* W. L. Byrne, ed., Molecular approaches to learning and memory. Academic Press, New York, pp. 285–284.

WOLTHUIS, O. L., J. E. ANTHONI and W. F. STEVENS, 1969, Acta Physiol. Pharmacol. Neerlandica *15*, 93.

ZEMP, J. W., J. E. WILSON, K. SCHLESINGER, W. O. BOGGAN and E. GLASSMAN, 1966, Proc. Natl. Acad. Sci. U.S.A. *55*, 1423.

ZIPPEL, H. P. and G. F. DOMAGK, 1969, Experientia *25*, 938.

Subject index

The biochemical and behavioral procedures listed in this index are in general those that have been used in transfer studies. Unless this could lead to confusion, the word transfer (in transfer; transfer of, and etc.) has been omitted under such entries. Similarly, since the large majority of the investigations reported here was made with rats and mice, no specific mention is made of these animals except where important to avoid misunderstanding. However, references are given to more unusual experimental subjects, e.g. fish.